Advances in Fluid Dynamics
with emphasis on
Multiphase and Complex Flow

WIT*PRESS*
WIT Press publishes leading books in Science and Technology.
Visit our website for the current list of titles.
www.witpress.com

WIT*eLibrary*
Home of the Transactions of the Wessex Institute.
Papers published in this volume are archived in the WIT elibrary in volume 132 of
WIT Transactions on Engineering Sciences (ISSN 1743-3533).
The WIT electronic-library provides the international scientific community with immediate
and permanent access to individual papers presented at WIT conferences.
http://library.witpress.com

ELEVENTH INTERNATIONAL CONFERENCE ON
INTERNATIONAL CONFERENCE ON ADVANCES IN FLUID DYNAMICS WITH
EMPHASIS ON MULTIPHASE AND COMPLEX FLOW

AFM/MPF 21

CONFERENCE CHAIRMEN

Santiago Hernández
University of A Coruña, Spain
Member of WIT Board of Directors

Peter Vorobieff
University of New Mexico, USA

INTERNATIONAL SCIENTIFIC ADVISORY COMMITTEE

ORGANISED BY

Wessex Institute, UK
University of New Mexico, USA

SPONSORED BY

WIT Transactions on Engineering Sciences
*International Journal of Computational Methods
and Experimental Measurements*

WIT Transactions

Wessex Institute
Ashurst Lodge, Ashurst
Southampton SO40 7AA, UK

Advances in Fluid Dynamics
with emphasis on
Multiphase and Complex Flow

Editors

Santiago Hernández
University of A Coruña, Spain
Member of WIT Board of Directors

Peter Vorobieff
University of New Mexico, USA

WIT_PRESS_ Southampton, Boston

Editors:

S. Hernández
University of A Coruña, Spain
Member of WIT Board of Directors

Peter Vorobieff
University of New Mexico, USA

Published by

WIT Press
Ashurst Lodge, Ashurst, Southampton, SO40 7AA, UK
Tel: 44 (0) 238 029 3223; Fax: 44 (0) 238 029 2853
E-Mail: witpress@witpress.com
http://www.witpress.com

For USA, Canada and Mexico

Computational Mechanics International Inc
25 Bridge Street, Billerica, MA 01821, USA
Tel: 978 667 5841; Fax: 978 667 7582
E-Mail: infousa@witpress.com
http://www.witpress.com

British Library Cataloguing-in-Publication Data

A Catalogue record for this book is available
from the British Library

ISBN: 978-1-78466-435-0
eISBN: 978-1-78466-436-7
ISSN: (print) 1746-4471
ISSN: (on-line) 1743-3533

Preface

This volume contains a selection of the papers presented at the 11th International Conference on Advances in Fluid Dynamics with emphasis on Multiphase and Complex Flow organised by the Wessex Institute (WIT), UK. This conference merged two conference series, the International Conference on Advances in Fluid Mechanics (AFM) and the International Conference on Computational and Experimental Methods in Multiphase and Complex Flow (MPF), which originated in New Orleans (1996), and Orlando, Florida (2001) respectively. While the primary motivation behind merging these events and holding the conference online was the well-being of delegates during the COVID-19 pandemic, in retrospective it makes perfect sense from the scientific point of view: modern fluid mechanics deals extensively with multiphase flow phenomena, which are ubiquitous in Nature and in engineering applications.

The complexity of the physics governing fluid flow (and multiphase flow specifically) necessitates a multi-pronged approach to most problems of interest, combining advancements in experiment, theory, and numerics within a robust validation, verification, and uncertainty quantification (VVUQ) framework. Papers presented at the conference described experiments characterizing single- and multiphase media and presenting important diagnostic developments such as application of machine learning to multiphase flow regime detection with a gamma-ray flowmeter, characterization of void fraction in gas-liquid mixtures with conductance probes, and a novel optical diagnostic for accurate particle distribution measurement. One experimental technique deserves special mention because it was developed for the extremely hostile environment inside the Fukushima Dai-ichi nuclear power plant damaged by the March 2011 earthquake and tsunami. Robots inspecting the plant interior can now search for leaks of contaminated water using an innovative ultrasonic velocity profiler (UVP).

Important theoretical advancements were also presented at the conference, such as a new formulation describing sedimentation in monodisperse suspensions, and improved constitutive relations for shock-compressed monatomic gases. Conference delegates described numerical studies using a variety of approaches and codes, but careful attention to VVUQ was the common theme. Another notable common feature shared by many of the papers was utilization of open-source software, both well-established (OpenFOAM) and new (FIESTA).

The conference brought renowned experts in the field together with young researchers from many countries. The format chosen for the event encouraged real-time online discussion of the conference presentations. Papers selected for this volume should attract further attention from researchers working in the field of fluid mechanics. As for all Wessex Institute conference publications, these papers are referenced by CrossRef. Publications of the Wessex Institute, including the WIT Transactions on Engineering Sciences (the present volume is a part of this series), are widely distributed throughout the world in digital as well as paper format. The electronic versions of all the papers are archived in the elibrary of the Wessex Institute (http://www.witpress.com/elibrary) where they are easily available to the scientific community.

We express our gratitude to the members of the Board, to the International Scientific Advisory Community of the conference, and to our colleagues who participated in the peer review of the contributions to this volume. The editors specially thank the authors featured in this volume for providing high-quality, thought-provoking and innovative papers.

The Editors, 2021

Contents

Section 3: Computational methods

Section 4: Theoretical and computational formulations

SECTION 1
MULTIPHASE FLOW

THREE-DIMENSIONAL VALIDATION EXERCISE FOR FIESTA CODE: EVOLUTION OF SHOCK-DRIVEN INSTABILITIES

BRIAN ROMERO[1,2], PETER VOROBIEFF[1], SVETLANA V. POROSEVA[1] & JON M. REISNER[2]
[1]University of New Mexico, USA
[2]Los Alamos National Laboratory, USA

ABSTRACT

In this paper, we present simulation results for the three-dimensional, shock-driven Kelvin–Helmholtz instability. Simulations are performed with a Mach 2.0 shock propagating through a finite-diameter cylindrical column of dense gas inclined at an angle θ with respect to the shock plane. After passage of the shock, the gas curtain has accelerated along its axis and a Kelvin–Helmholtz instability forms on the column surface. This is the first known numerical reproduction of this phenomena in three dimensions, which has previously been observed in experiments with an inclined cylindrical gas column. The effects of changes to initial column angle ($\theta = 0°$, $10°$, $20°$, $30°$) are explored in detail to complement experimental data. The effects of shock reflection near the base of the column are also examined to identify a possible flow perturbation near the foot which was seen in our previous two-dimensional numerical studies of shock-accelerated inclined gas curtains. The overall flow morphology compares well with experimental data in a cross-sectional plane through the column midpoint and a vertical plane through the column axis. Simulations were performed with FIESTA, an exascale ready, GPU accelerated compressible flow solver developed at the University of New Mexico.
Keywords: multiphase flow, mixing, modeling, compressible flow, validation.

1 INTRODUCTION

In many problems of fundamental or practical interest, transition to turbulence is driven by a shock propagating through a non-uniform medium, such as gases of different densities, possibly with non-gaseous inclusions in the form of particles or liquid drops. In astrophysics, shocks from supernova explosions and asymptotic giant branch stars propagate through gas and cosmic dust [1]–[3], leading to formation of structures on the scale of light years [4], [5]. In determining the origin of cosmic dust, a factor of great interest is its evaporation in supernovae [6]. In geophysics, shock waves and impulsive acceleration events that strongly affect Earth's magnetosphere occur as the result of interaction between the solar wind and Earth's magnetopause [7]. Engineering applications include fuel mixing in scramjets [8], explosion physics [9], [10], and inertial confinement fusion [11]. The mixing of liquid fuel droplets with air by oblique shock waves creates issues for scramjet engines [12]. An understanding of detonation waves with a multiphase air-fuel droplet mixture is essential for designing more practical liquid fueled pulse detonation and rotating detonation engines [13].

There are three well-known interfacial instabilities: Kelvin–Helmholtz (KHI), driven by shear, Rayleigh–Taylor (RTI), driven by sustained acceleration in the direction of decreasing density across the interface, and Richtmyer–Meshkov (RMI), driven by impulsive acceleration across the interface. It is possible to set up a problem where one of these instabilities would clearly dominate, but in many situations of practical interest, all three may play some role. The situation is further complicated if the medium of interest is multiphase. In the simpler case, a multiphase medium would be comprised of a single phase occupying a dominant volume fraction, with small inclusions of the second phase (droplets or particles in gas, bubbles in fluid), as even a small volume fraction can have a large fraction of the total mass. Droplets or particles carried by the gas flow strongly change the relevant physics, e.g.,

WIT Transactions on Engineering Sciences, Vol 132, © 2021 WIT Press
www.witpress.com, ISSN 1743-3533 (on-line)
doi:10.2495/MPF210011

by changing the compressibility of the medium (fluid with bubbles) or its average density. Moreover, for the case of shock-accelerated gas carrying particles or droplets, it cannot be safely assumed that the inclusions remain embedded in the post-shock flow. As the particles increase in size they become slow to react and the length and time scales of the effects of their interaction with the surrounding medium must be carefully considered. For droplets, break-up or coagulation may be precipitated by the shock passage. The addition of phase change and reactions also influences the evolution of the flow greatly and brings additional complexity to the problem.

Numerical modeling of shock-driven compressible flow, especially when the medium is multispecies and/or multiphase, remains a challenging problem which cannot be solved with a brute-force approach despite the recent advances in computational power. The range of scales that need to be resolved would be prohibitively computationally expensive – from molecular mean free path (representative value 60 nm for nitrogen at atmospheric pressure) to maximum system scale (which could be light-years for astrophysical problems). Within this range spanning multiple orders of magnitude, different physics would manifest at different scales: the shock front thickness is on the same order as the mean free paths, the Kolmogorov microscales at and below which viscous dissipation becomes dominant are typically in the micron range, the scales associated with particle/droplet interaction are commensurate with the particle/droplet sizes, and so on. Accordingly, in most cases the computational scheme is organized in some way that does not directly resolve the smallest scales – commonly by time-averaging small-scale fluctuations, leading to Reynolds-averaged Navier–Stokes equations (RANS), or by low-pass filtering of the Navier–Stokes equations, resulting in LES (large-eddy simulation) computational schemes. In both cases, additional conjectures are required to account for extra variables introduced due to averaging (RANS, closure models) or to model subgrid scales (LES).

Another approach is modeling full Navier–Stokes (or Euler, if viscosity is to be disregarded) equations without resolving all the scales, but with no subgrid model [14]. This approach is known as implicit LES (ILES). Here we present a validation exercise for a newly developed, open-source code modeling Euler equations and resolving the flow structure on scales about an order of magnitude larger that the Kolmogorov microscale. The goal of the study is twofold: confirm that the code faithfully reproduces experimental results and gain insights into the flow by analyzing numerically modeled features that cannot be tracked with available experimental diagnostics.

2 EXPERIMENTAL ARRANGEMENT AND VALIDATION DATA

The experiments were conducted at the University of New Mexico tiltable shock tube [15]–[17]. Prior to release of the shock, the optically transparent test section of the shock tube is maintained at atmospheric pressure. A cylindrical jet of heavy gas (a mixture of sulfur hexafluoride SF_6 and acetone with traces of air) is injected downward through the top wall of the test section and exits through a hole in the bottom wall. This jet is surrounded with a co-flowing cylindrical jet of air, so that the flow is laminar and any mixing between the air (density ρ_1) and heavy gas (density ρ_2) is diffusive. The entire shock tube can be tilted to an angle θ with respect to the horizontal. With a column injection system chosen for a specific θ, the jet with nominal diameter D_{IC} flows as shown in Fig. 1.

Once these initial conditions with a cylindrical diffuse density interface are established, the shock tube produces a planar shock wave at a prescribed Mach number M. The shock front propagates from the driver section (not shown in Fig. 1) toward the test section. Behind the shock front the air is compressed and moving with piston velocity ΔV.

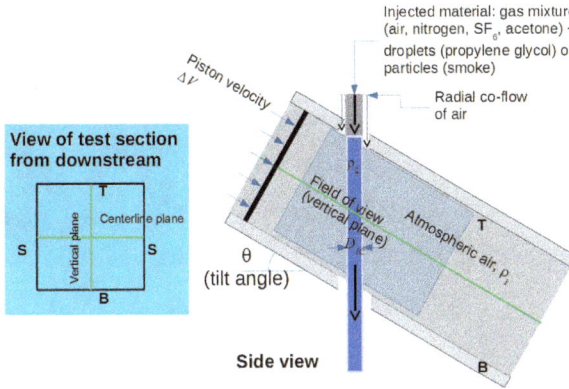

Figure 1: Schematic of the experimental arrangement for studies of shock interaction with an inclined cylindrical density interface (same setup as [16], [17]). Walls of the test section of the shock tube (and the corresponding computational domain) are labeled as T – top, B – bottom, and S – side.

During the experiments, pressure traces are collected at multiple downstream locations to monitor the shock propagation and trigger the imaging system, which can capture images in two planes. The first visualization plane (vertical) is parallel to the shock direction and equidistant from the vertical walls of the test section. The second plane (centerline) is similarly parallel to the top and bottom walls and tilted at the same angle θ as the shock tube. Each plane can be illuminated with precisely timed, short ($\sim$ 5 ns) pulses of a frequency-quadrupled (wavelength 266 nm) Nd:YAG laser. The laser beam passes through a combination of a cylindrical and a spherical lens, forming a laser sheet that selectively illuminates a narrow (submillimeter) planar cross-section of the flow. The acetone in the injected heavy gas mixture fluoresces in the visible range at 480 nm when lit with the 266-nm laser pulse.

The governing parameters of the flow are D_{IC}, θ, M and the Atwood number

$$A = \frac{\rho_2 - \rho_1}{\rho_2 + \rho_1}$$

Fig. 2 shows an example of an image sequence in two planes. Dimensionless time τ is defined there in accordance with Richtmyer's linear theory as

$$\tau = kA\Delta Vt,$$

where t is dimensional time and $k = 2\pi/D_{IC}$ is the dominant wavenumber of the initial density perturbation. In terms of τ, the initial RMI growth rates should remain the same for the same initial condition geometry, not changing with A or M (although this assertion is only valid for very early – linear – perturbation growth). The volume in the flow where the perturbed heavy gas mixes with the surrounding lighter gas is commonly referred to as the mixing zone, and its streamwise and spanwise extent can serve as quantitative code-validation benchmarks.

In the centerline plane, the dominant flow feature is a pair of counter-rotating vortices that roll up due to RMI. In the vertical plane, however, small-scale vortices form on the leading and trailing edges of the heavy-gas cylinder [16]. Based on the observed morphology,

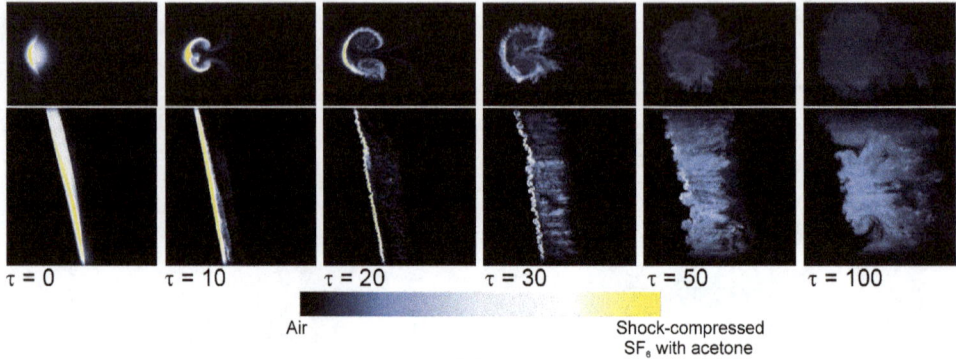

Figure 2: Representative experimental image sequence showing shock acceleration of a gas cylinder at a tilt angle $\theta = 20°$, $M = 2$, $A = 0.6$ [17]. τ is dimensionless time (refer to text). The top image for each τ shows the centerline plane, the bottom image – the vertical plane. Images are shown in false color to emphasize feature formation. Fluorescence intensity (color bar) increases monotonically with acetone (heavy gas tracer) concentration. Streamwise extent of the images is 44 mm, the field of view is following the mixing zone.

it was suggested that the vortex formation mechanism responsible for these features in the vertical plane is KHI due to shock-driven vorticity deposition on the inclined density interfaces [16], [18], [19]. As the flow evolves, secondary instabilities begin to emerge in both planes, leading to enhanced mixing in increasingly disordered flow that at late time manifests statistics consistent with turbulence [17].

The experimental results described above can be used to develop a well-characterized set of quantitative benchmarks for validation of numerical codes modeling shock-driven mixing and /or multphase flow. We begin with a validation exercise for the case of a single-phase density interface, as described in the next section.

3 NUMERICAL SETUP

The computational framework for simulations is FIESTA (Fast Interface Evolution, Shocks and Transition in the Atmosphere). It is a new open-source computational fluid dynamics code developed at the University of New Mexico to study instabilities and transition to turbulence. FIESTA is capable of carrying out simulations at scales ranging from those characteristic for a laboratory shock tube to large atmospheric scales.

The code is written in the modern C++ language and takes advantage of object-oriented techniques to improve the code modularity. FIESTA uses the Kokkos C++ Performance Portability Ecosystem [20] to target both traditional CPU architectures and General Purpose Graphics Processing Units (GPGPUs) while avoiding the code complexity and duplication commonly encountered when supporting different architectures. Output files are formatted using the HDF5 file format [21]. Input files are written in the Lua language [22] allowing for easy parameterization of a problem. Multi-GPGPU support is provided by the Message Passing Interface (MPI).

The modular code design allows to include different physics and models as needed. Existing modules can solve two-dimensional (2D) and three-dimensional (3D) two-species Euler and NS equations on generalized curvilinear or uniform Cartesian structured grids. The Euler equations are solved with a simplified 5th-order weighted essentially non-oscillatory

(WENO5) finite difference scheme [23], [24]. The time scheme used is a low storage, second-order, explicit Runge–Kutta integrator. At each time level, advection terms are approximated using the WENO5 scheme. The pressure gradient term is approximated using a fourth-order central difference scheme [23], [24].

To obtain the results presented here, we used fully three-dimensional, two-species, inviscid Euler equations in their conservative form. These equations consist of the continuity equations for each gas species, equations for each momentum component, and the conservation equation for specific total energy [25]. The size of the computational domain was $40 \times 10 \times 5$ cm in the x (streamwise), y (vertical) and z (normal to the $x - y$ plane) directions respectively. The equations were solved on a uniform rectangular grid: $dx = dy = dz = 67$ μm, with reflective conditions on the top and bottom of the domain, and outflow conditions upstream, downstream and on each side in the $z-$direction. Unlike experiment, where the initial conditions are formed by gravity-driven flow, gravity is not considered, instead the heavy-gas column is tilted to an angle θ in the computational domain. The column is modeled as initially diffusive, with gas concentration profiles matching experimental measurements [16]. Note that in experiments, the gravity-driven jet of heavy gas has a velocity on the order of centimeters per second prior to shock acceleration – this is disregarded in the modeling, as the representative piston velocity of the shock-accelerated flow at $M = 2$ is 434 m/s.

Dalton's law is assumed to govern the pressure of gas mixtures, and gases are modeled with ideal-gas equation of state, with gas constants $\gamma = 1.402$ for air and $\gamma = 1.095$ for injected heavy gas (gas constant for SF_6). The pressure and temperature in the gases prior to shock arrival are set at 300 K and 1 bar.

4 RESULTS AND DISCUSSION

The work presented here is a step forward from the implementation of the 2D module of FIESTA [25], which was used to simulate shock interaction with an inclined gas curtain (not cylinder) and to confirm the two-dimensional inviscid model of shock-driven KHI originally based on experiment [16]. Fig. 3 shows a time sequence of density maps in two planes obtained as a result of the three-dimensional FIESTA run simulating the experiment whose initial conditions are schematically shown in Fig. 1 and results – in Fig. 2. There is good agreement between simulation and experiment – both qualitative and quantitative. The flow features observed in experiment are faithfully reproduced, including the counter-rotating vortices in the centerline plane, the Kelvin–Helmholtz vortices in the vertical plane, and the emergence of secondary instabilities.

Quantitative comparison of experiments and numerics is still ongoing, but preliminary measurements show good agreement in terms of integral flow features such as mixing zone width (extent of the perturbed column in the streamwise direction) and the dominant wavelength of the KHI in the vertical plane [25].

Similar observations can be made with regard to the $\theta = 30°$ simulation, Fig. 4. However, the computational exercise we present here goes beyond experimental validation of the 3D FIESTA module. The diagnostics we use in experiment (local pressure measurements at four points and planar laser-induced Fluorescence – PLIF – imaging in two planes) cannot reveal some important flow features, for example, the 3D geometry of the shock reflected from the inclined gas column ($\tau = 20$ in Fig. 3 and $\tau = 14$ in Fig. 4). This pattern is remarkably similar to shock reflection off an elongated solid body, which could offer some hints about the influence of the Mach number on the nonlinear instability growth: it was shown that growth plots for different initial conditions and Mach numbers can be collapsed to a single curve in dimensionless coordinates with time τ and the instability amplitude nondimensionalized with

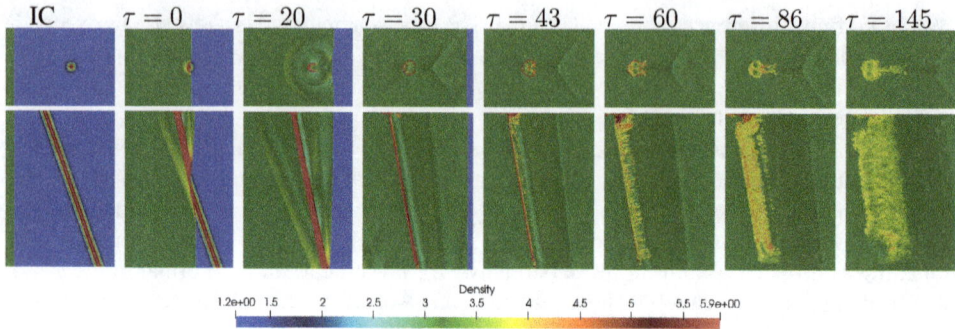

Figure 3: Density maps in the centerline (top row) and vertical (bottom row) planes obtained with FIESTA, $M = 2$, $A = 0.6$, $\theta = 30°$. Density units (color bar) are kg/m^3. Only a part of the streamwise extent of the computational domain with features of interest is shown, similar to Fig. 2. First image (label "IC") shows the initial conditions. $\tau = 0$ corresponds to the time when the shock arrives at the column in the centerline plane.

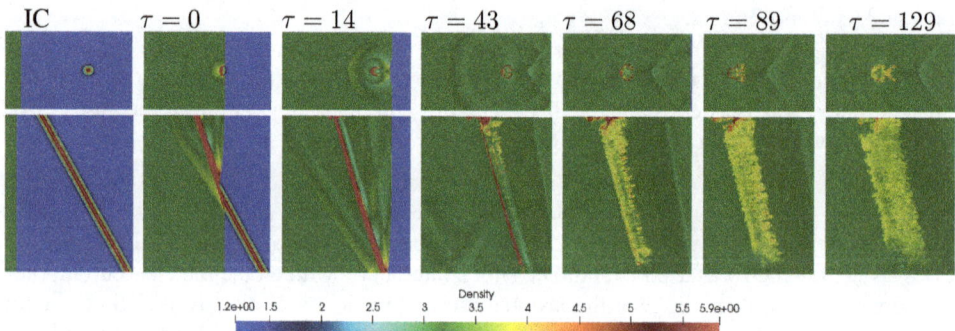

Figure 4: Density maps in the centerline (top row) and vertical (bottom row) planes obtained with FIESTA, $M = 2$, $A = 0.6$, $\theta = 30°$. Density units (color bar) are kg/m^3. Only a part of the streamwise extent of the computational domain with features of interest is shown. [16], [18], contain corresponding experimental images and data for validation.

a scaling $w_0 M^\kappa$, where w_0 is the minimal perturbation amplitude after shock compression and phase inversion, and $\kappa \sim 0.5$ [16], [26]. A similar $M^{0.5}$ power-law scaling is known from ballistics [27], [28], so it was conjectured that similar wake-like features in the flows are responsible for the scalings. The simulation results shown here confirm the presence of such features.

While focus on the mixing zone features (Figs 2–4) is important for experimental validation of the code, it is also instructive to examine the entirety of the computational domain. Fig. 5 shows density maps revealing several important features beyond the mixing zone. Reflected waves from shock-column interaction propagate upstream. In 2D simulations of a gas curtain [25], surface-shock interaction near the foot of the inclined curtain was shown to produce a length scale that is amplified by shock-driven KHI. A similar phenomenon is now apparent in the 3D simulation. Downstream, a spike of dense material ejected from the cylinder moves with a high subsonic velocity ($\sim 0.3M$ in terms of the local Mach

$\tau = 145, \theta = 20°$

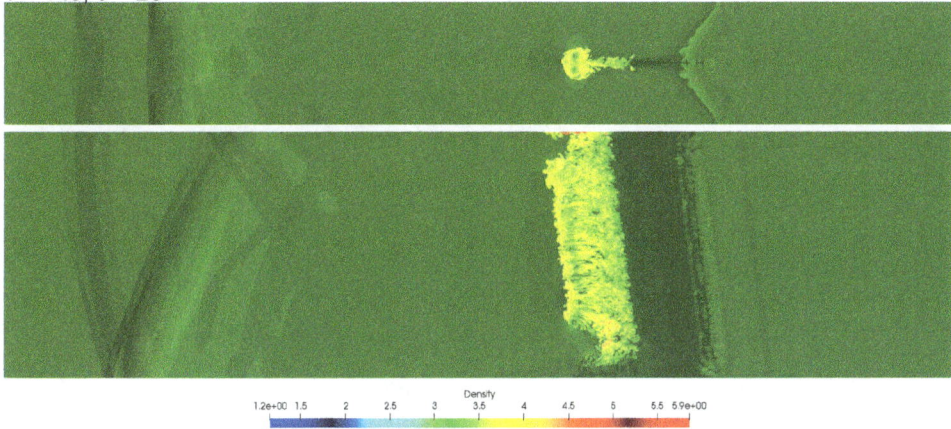

Figure 5: Centerline and vertical plane density maps spanning the entire computational domain at late times. Density units (color bar) are kg/m^3, palette is different from the one used in Figs 3 and 4 to emphasize features outside the mixing zone, including the material ejected from the column in the downstream direction due to shock focusing and pressure wave reflection propagating upstream.

number [29]). Vortex structures at the end of the spike have been observed in experiment ("lion's tail" [29]), however, the adjacent wakelike structure best visible in the centerline plane of Fig. 5 could not be visualized because there was no fluorescent tracer to track it.

5 CONCLUSIONS AND FUTURE WORK

Our numerical studies of three-dimensional interaction between a planar shock front and an inclined, initially diffuse column of heavy gas help validate the 3D Euler equations module of the new open-source FIESTA code. Additionally, the modeling reveals important flow features that would be difficult to resolve experimentally with fluorescence-based flow visualization techniques.

Follow-up work with FIESTA will include grid convergence studies, an implementation of a compressible multiphase flow module, and incorporation of different equations of state for single gases and mixtures to account for real-gas effects.

FIESTA validation exercises will be extended beyond integral quantitative benchmarks, such as streamwise mixing zone extent, to robust statistical benchmarks based on velocity and scalar structure function scalings [17], [30], [31], fractal characteristics of the mixing interfaces [32], and species concentration statistics [16].

ACKNOWLEDGEMENTS

Peter Vorobieff is a Halliburton Professor. The Halliburton Professorship was established in 1982 by the Halliburton Foundation, and the endowment was subsequently expanded by the addition of funds from the State of New Mexico. This research is partially supported by the Defense Threat Reduction Agency (DTRA) grant HDTRA-18-1-0022. Mr. Romero acknowledges support by the New Mexico Consortium. We also wish to thank the UNM Center for Advanced Research Computing, supported in part by the US National Science Foundation, for providing some of the high performance computing resources used in this work.

REFERENCES

[1] Mendis, D.A. & Rosenberg, M., Cosmic dusty plasma. *Annual Review of Astronomy and Astrophysics*, **32**(1), pp. 419–463, 1994.

[2] Bocchio, M., Jones, A.P. & Slavin, J.D., A re-evaluation of dust processing in supernova shock waves. *Astronomy & Astrophysics*, **570**, p. A32, 2014.

[3] Woitke, P., 2D models for dust-driven AGB star winds. *Astronomy & Astrophysics*, **452**(2), pp. 537–549, 2006.

[4] Chevalier, R.A., Blondin, J.M. & Emmering, R.T., Hydrodynamic instabilities in supernova remnants – Self-similar driven waves. *The Astrophysical Journal*, **392**, pp. 118–130, 1992.

[5] Kane, J., Drake, R. & Remington, B., An evaluation of the Richtmyer-Meshkov instability in supernova remnant formation. *The Astrophysical Journal*, **511**(1), p. 335, 1999.

[6] Bianchi, S. & Schneider, R., Dust formation and survival in supernova ejecta. *Monthly Notices of the Royal Astronomical Society*, **378**(3), pp. 973–982, 2007.

[7] Wu, C., Shock wave interaction with the magnetopause. *Journal of Geophysical Research: Space Physics*, **105**(A4), pp. 7533–7543, 2000.

[8] Yang, J., Kubota, T. & Zukoski, E.E., Applications of shock-induced mixing to supersonic combustion. *AIAA Journal*, **31**(5), pp. 854–862, 1993.

[9] DOE ASCI teraflops computer fully running for first time. Sandia National Laboratories News Release, 1997. http://www.sandia.gov/media/online.htm.

[10] Benjamin, R.F., An experimenter's perspective on validating codes and models with experiments having shock-accelerated fluid interfaces. *Computing in Science & Engineering*, **6**(5), pp. 40–49, 2004.

[11] Goncharov, V., Theory of the ablative Richtmyer-Meshkov instability. *Physical Review Letters*, **82**(10), p. 2091, 1999.

[12] Pandey, K. & Sivasakthivel, T., Recent advances in scramjet fuel injection – A review. *International Journal of Chemical Engineering and Applications*, **1**(4), p. 294, 2010.

[13] Huang, Y., Tang, H., Li, J. & Zhang, C., Studies of DDT enhancement approaches for kerosene-fueled small-scale pulse detonation engines applications. *Shock Waves*, **22**(6), pp. 615–625, 2012.

[14] Margolin, L.G., Rider, W.J. & Grinstein, F.F., Modeling turbulent flow with implicit LES. *Journal of Turbulence*, (7), p. N15, 2006.

[15] Anderson, M., Vorobieff, P., Truman, C., Corbin, C., Kuehner, G., Wayne, P., Conroy, J., White, R. & Kumar, S., An experimental and numerical study of shock interaction with a gas column seeded with droplets. *Shock Waves*, **25**(2), pp. 107–125, 2015.

[16] Olmstead, D., Wayne, P., Yoo, J.H., Kumar, S., Truman, C.R. & Vorobieff, P., Experimental study of shock-accelerated inclined heavy gas cylinder. *Experiments in Fluids*, **58**(6), p. 71, 2017.

[17] Olmstead, D., Wayne, P., Simons, D., Monje, I.T., Yoo, J.H., Kumar, S., Truman, C.R. & Vorobieff, P., Shock-driven transition to turbulence: Emergence of power-law scaling. *Physical Review Fluids*, **2**(5), p. 052601, 2017.

[18] Wayne, P., Olmstead, D., Vorobieff, P., Truman, C. & Kumar, S., Oblique shock interaction with a cylindrical density interface. *WIT Transactions on Engineering Sciences*, vol. 89, WIT Press: Southampton and Boston, pp. 161–169, 2015.

[19] Romero, B.E., Poroseva, S., Vorobieff, P. & Reisner, J., Shock driven Kelvin-Helmholtz instability. *AIAA Scitech 2021 Forum*, p. 0051, 2021.

[20] Edwards, H.C., Trott, C.R. & Sunderland, D., Kokkos: Enabling manycore performance portability through polymorphic memory access patterns. *Journal of Parallel and Distributed Computing*, **74**(12), pp. 3202–3216, 2014.

[21] Koranne, S., Hierarchical data format 5: HDF5. *Handbook of Open Source Tools*, Springer, pp. 191–200, 2011.

[22] Ierusalimschy, R., *Programming in Lua*, Roberto Ierusalimschy, 2006.

[23] Ramani, R., Reisner, J. & Shkoller, S., A space-time smooth artificial viscosity method with wavelet noise indicator and shock collision scheme, Part 1: The 1-D case. *Journal of Computational Physics*, **387**, pp. 81–116, 2019.

[24] Ramani, R., Reisner, J. & Shkoller, S., A space-time smooth artificial viscosity method with wavelet noise indicator and shock collision scheme, Part 2: The 2-D case. *Journal of Computational Physics*, **387**, pp. 45–80, 2019.

[25] Romero, B., Poroseva, S., Vorobieff, P. & Reisner, J., Simulations of the shock driven Kelvin-Helmholtz instability in inclined gas curtains. *Physics of Fluids*, 2021. In press.

[26] Orlicz, G., Balasubramanian, S. & Prestridge, K., Incident shock mach number effects on Richtmyer-Meshkov mixing in a heavy gas layer. *Physics of Fluids*, **25**(11), p. 114101, 2013.

[27] McCoy, R., *Modern Exterior Ballistics: The Launch and Flight Dynamics of Symmetric Projectiles*, Schiffer Pub., 1999.

[28] Roetzel, W., Analytical calculation of trajectories using a power law for the drag coefficient variation with Mach number. *WIT Transactions on Modelling and Simulation*, vol. 40, WIT Press: Southampton and Boston, 2005.

[29] Bernard, T., Randall Truman, C., Vorobieff, P., Corbin, C., Wayne, P.J., Kuehner, G., Anderson, M. & Kumar, S., Observation of the development of secondary features in a Richtmyer–Meshkov instability driven flow. *Journal of Fluids Engineering*, **137**(1), 2015.

[30] Vorobieff, P., Rightley, P.M. & Benjamin, R.F., Power-law spectra of incipient gas-curtain turbulence. *Physical Review Letters*, **81**(11), p. 2240, 1998.

[31] Vorobieff, P., Mohamed, N.G., Tomkins, C., Goodenough, C., Marr-Lyon, M. & Benjamin, R., Scaling evolution in shock-induced transition to turbulence. *Physical Review E*, **68**(6), p. 065301, 2003.

[32] Vorobieff, P., Rightley, P.M. & Benjamin, R.F., Shock-driven gas curtain: Fractal dimension evolution in transition to turbulence. *Physica D: Nonlinear Phenomena*, **133**(1–4), pp. 469–476, 1999.

DESIGN OF OPTIMAL CONDUCTANCE PROBES FOR TWO-PHASE FLOW TOMOGRAPHY AND LIQUID HOLDUP: APPLICATION TO THE DETERMINATION OF THE AVERAGE VOID FRACTION IN A REGION

JOSÉ LUIS MUÑOZ-COBO, YAGO RIVERA, CÉSAR BERNA & ALBERTO ESCRIVÁ
Instituto de Ingeniería Energética, Universitat Politècnica de València (UPV), Spain

ABSTRACT

The knowledge of the characteristics of two-phase flows is an important issue in a wide variety of engineering applications. The design of optimal conductance probes for two-phase tomography allows the determination of the shape of a rapidly moving wavy film of water under gravity and shear stress forces exerted on the interface by a gas-phase. Obtaining the characteristics of the interfacial waves can be achieved with a proper design of this kind of sensors. Other application of these probes is for holdup, in this case the determination by conductance measurements of the average void fraction inside a region depends on the effective conductivity of the mixture, which depends not only on the void fraction but also on its distribution inside the region, which in general relies on the two-phase flow regime. In this paper we use the expressions developed recently for the relative conductance of two-plate conductance sensors and those developed previously for two-ring sensors, in order to obtain the best probe designs to be used in different applications of annular two-phase flow tomography. In addition, we discuss the best way to optimize these probes to achieve better results and accuracy when measuring the characteristics of the disturbance and ripple waves as height, frequency and amplitude produced in annular two-phase flow. Also, we use these expressions to optimize the design of the sensors for hold up applications, discussing the effect of different void fraction distributions on the effective conductivity and the measured average void fraction. The computed results have been validated with experimental data obtained from different well-known authors in this field. Finally, we discuss the best type of conductance probes to be used for each specific application and the geometric characteristics of its electrodes as width, height, and separation between them.

Keywords: two-phase sensor tomography, interfacial wave measurements, holdup, plate conductance probes, sensor design, void fraction determination.

1 INTRODUCTION

The knowledge of the characteristics of two-phase flow regimes is an important issue in many two-phase flows devices currently used in the energy and chemical industries [1]–[4]. There are several two-phase flow patterns in horizontal, and vertical pipes, being the annular flow the most relevant one in steam generators of energy plants (nuclear power and combined cycles), refrigeration systems, and pipes used for the transport and production of natural gas [4]–[6]. Each flow pattern in two-phase flow corresponds to a characteristic gas-liquid topology, which defines typical interfaces between the phases of a given flow, as bubbly flow, annular flow, or cap/slug flow. The exact knowledge of the structures as disturbance and ripple waves formed at the interface in annular flow is important for several reasons, the first one is that the heat transfer between the phases depends on the shape and height of these waves, the second is that the pressure drop is highly influenced by these structures. In addition, the third reason is that when the gas phase moves at higher velocity that the liquid phase the shear stress produced at the interface can tear off small drops from the peaks of the disturbance waves reducing its height [7]. Therefore, as observed experimentally [7], [8], over a wide range of liquid and gas superficial velocities the two-phase annular flow pattern in a pipe consists of a thin liquid film moving adjacent to the wall, with a gaseous core

WIT Transactions on Engineering Sciences, Vol 132, © 2021 WIT Press
www.witpress.com, ISSN 1743-3533 (on-line)
doi:10.2495/MPF210021

transporting small drops entrained from the crest of the wavy film. In addition, if the velocity of the gas is smaller than the threshold for drop-entrainment then the drops are not present in the gaseous core.

Several methods have been used in the past to obtain the two-phase flow characteristics, as void fraction distribution in bubbly and cap/slug flow, or wave characteristics in annular flow. To obtain void fraction distribution the most common technique is by conductivity probes with two and three needles [9], [10], and mesh wire sensors [11], while to obtain the average void fraction inside a region one can use the conductance probe method [1], the impedance method [12], [13], and the radiation attenuation method [14]. Finally, to obtain the characteristics of the interfacial waves in annular flow the most common methods are the conductance methods, [1], [15] and the optical ones as the optical laser-based measurement technique, also known as Planar Laser-Induced Fluorescence (PLIF) [16].

The main goal of this paper is to describe how to obtain the best design of the conductance probes, which are used for two-phase tomography mainly of annular flows and holdup applications as determining the average liquid fraction or the average void faction inside a pipe containing a two-phase flow mixture.

The paper has been organized as follows, Section 2 introduces the conductance probe method, presenting in Section 2.1 the expressions for the calculation of the relative conductance in ring and plate conductance sensors. In Section 2.2 we show the expressions used for holdup applications using two-ring electrodes. Then, in Section 2.3 we show how to obtain the error factor used for the design of conductance probes, and the useful range of film thicknesses that we can measure with a conductance probe. Sections 3.1 and 3.2 are devoted to the comparison of the results obtained with different sensor designs of plate probes, while in Section 3.3 we study the use of the conductance sensors for holdup and void fraction determination. Finally, in Section 4 we give the main conclusions of the paper.

2 THE CONDUCTANCE PROBE METHOD

In this section we describe the main characteristics of the conductance probe method. It is well known that the measured electrical impedance between a pair of electrodes submerged in a conducting fluid as tap water is essentially resistive when the frequency of the exciting a.c. signal of the emitting electrode is in the range from 10 to 100 kHz, in this range of frequencies the impedance methods are known as conductance methods. Two types of electrodes, ring, and plate have been used in conductance probes to measure the film thickness of the conducting fluid [2], [15], [17]. In this section we will study the performance of flush mounted electrodes to determine the best probe designs to have good spatial resolution and sensitivity to perform measurement of the film thickness.

2.1 The conductance probe method with plate and ring electrodes

The two-ring conductance probe consist of two ring shape electrodes mounted along the inner circumference of the pipe, this type of electrodes has been studied by Tsochatzidis et al. [2] and Fossa [17]. This kind of conductance probes provides good results to obtain the void fraction in uniformly dispersed bubbly flow regime and even in non-homogeneous bubbly flow as discussed recently by Muñoz-Cobo et al. [1]. The two-plate conductance probe consists in two flush mounted plated electrodes, subjected to a high frequency a.c. electrical excitation with frequencies ranging from 10 to 100 kHz, at these frequencies the electric impedance through the conducting liquid is essentially resistive. To calibrate this type of electrodes, to measure film thickness, a set of dielectric cylinders of different radius R_{in} are mounted inside the pipe as displayed at Fig. 1(a). The probe conductance defined as the ratio

$G = I/\delta\phi$, is then measured being I the electric intensity circulating through the conducting fluid (electrolyte), and $\delta\phi$ the average electric-potential difference between the emitter and receiver electrodes of the two-ring or two-plate probes i.e., $\delta\phi = \langle\phi_E\rangle - \langle\phi_R\rangle$, which depends on the film thickness δ of the conducting fluid. If this thickness is very small, then the conductance is also small because of the electric resistance to the pass of ions through the electrolyte is large. Normally, people use the non-dimensional conductance $G^* = G/(\sigma_w l)$, which is defined as the ratio of the conductance and the product of the conducting fluid conductivity σ_w and the characteristic length l of the sensor.

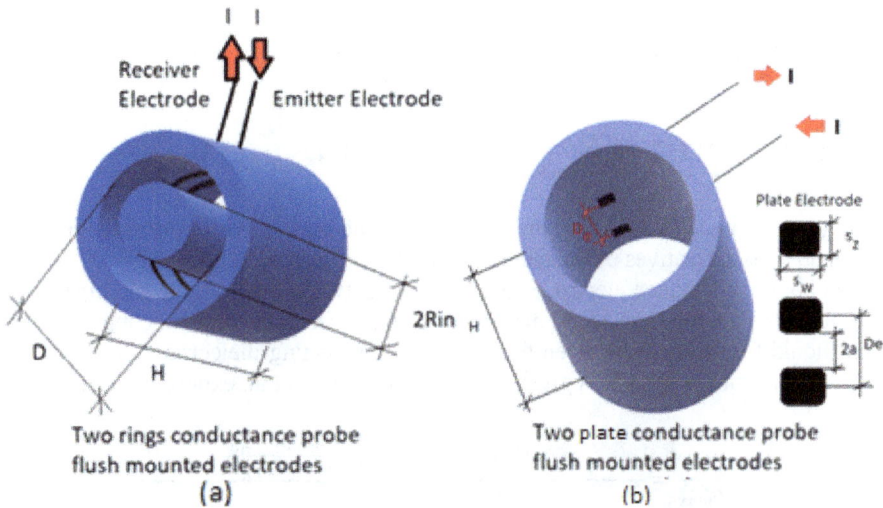

Figure 1: (a) Two-ring conductance probe with inner dielectric calibration cylinder; and (b) Two plate conductance probe displaying the geometrics characteristics of the electrodes.

The geometric characteristics of a two-electrode conductance sensor are defined by the following geometric lengths, $2a$, which is the distance between the electrodes, D_e that is the distance between the electrode centres, s_z that is the height of each electrode in the axial direction and s_w that is the width of the electrodes.

However, to present and compare the experiments of different authors with theory, it is convenient due to the variations of the conductivity from one experiment to another, to normalize the non-dimensional conductance G^* with the conductance obtained when the pipe is full of water G^*_{max}, in this way it is defined the relative conductance G^*_{rel}. If one assumes that the sensor probe is located at the centre of the pipe, then the relative conductance expressions for a two-ring conductance sensor obtained by Tsochatzidis et al. [1], [2], and for a two-plate conductance sensor deduced by Muñoz-Cobo et al. [2], can be unified into the following equation:

$$G^*_{rel} = \frac{G^*}{G^*_{max}} = \frac{C'_1 \sum_{n=0}^{\infty} \frac{b_n^2}{(2n+1)^3} \frac{I_0(\gamma_n R)}{I_1(\gamma_n R)} + C'_2 \sum_{n=0}^{\infty}\sum_{m=1}^{\infty} \frac{e_{m,n}^2}{(2n+1)^3 m} \frac{I_m(\gamma_n R)}{I'_m(\gamma_n R)}}{C'_1 \sum_{n=0}^{\infty} \frac{b_n^2}{(2n+1)^3} f(\gamma_n R_{in}, \gamma_n R) + C'_2 \sum_{n=0}^{\infty}\sum_{m=1}^{\infty} \frac{e_{m,n}^2}{(2n+1)^3 m} f_m(\gamma_n R_{in}, \gamma_n R)}, \tag{1}$$

being $C_1' = \frac{\Delta\theta^2}{4}$, $C_2' = \begin{cases} 1 \text{ for two} - \text{plate electrodes} \\ 0 \text{ for two} - \text{ring electrodes} \end{cases}$, $\gamma_n = \frac{(2n+1)\pi}{H}$, and $\Delta\theta = \frac{s_w}{R}$. In addition, we have defined the following quantities:

$$e_{m,n} = b_n \sin\left(\frac{m\Delta\theta}{2}\right) \text{ with } b_n = \cos\left(\gamma_n \frac{D_e + s_z}{2}\right) - \cos\left(\gamma_n \frac{D_e - s_z}{2}\right), \qquad (2)$$

where the functions $f(\gamma_n R_{in}, \gamma_n R)$, and $f_m(\gamma_n R_{in}, \gamma_n R)$, which appear in the denominator of eqn (1) are given by the expressions:

$$f(\gamma_n R_{in}, \gamma_n R) = \frac{I_0(\gamma_n R)}{I_1(\gamma_n R)} \left\{ \frac{1 + a_r(\gamma_n R_{in}) \frac{K_0(\gamma_n R)}{I_0(\gamma_n R)}}{1 - a_r(\gamma_n R_{in}) \frac{K_1(\gamma_n R)}{I_1(\gamma_n R)}} \right\} \text{ with } a_r(\gamma_n R_{in}) = \frac{I_1(\gamma_n R_{in})}{K_1(\gamma_n R_{in})}, \qquad (3)$$

$$f_m(\gamma_n R_{in}, \gamma_n R) = \frac{I_m(\gamma_n R) - a_{m,n}(\gamma_n R_{in}) K_m(\gamma_n R)}{I_m'(\gamma_n R) - a_{m,n}(\gamma_n R_{in}) K_m'(\gamma_n R)} \text{ with } a_{m,n}(\gamma_n R_{in}) = \frac{I_m'(\gamma_n R_{in})}{K_m'(\gamma_n R_{in})}, \qquad (4)$$

being $I_m(x), K_m(x), I_m'(x), K_m'(x)$, the first and second class modified Bessel functions of order m and their derivatives evaluated at $x = \gamma_n R_{in}$ and $x = \gamma_n R$.

When both conductance electrodes are parallels between them and to the pipe axis and flush mounted to the pipe surface as displayed at Fig. 2. Then, the relative conductance for a uniform liquid film located between the inner non conducting dielectric cylinder of radius R_{in} and the pipe inner radius R is given by the Muñoz-Cobo et al. expression [2]:

$$G_{rel}^* = \frac{G^*}{G_{max}^*} = \frac{C_1 \sum_{m=1}^{\infty} \frac{a_m^2}{m^3} + \sum_{n=1}^{\infty}\sum_{m=1}^{\infty} \frac{c_{m,n}^2}{m^2 n^3} \frac{I_m(\gamma_n' R)}{I_m'(\gamma_n' R)}}{C_1 \sum_{m=1}^{\infty} \frac{a_m^2}{m^3} \left(\frac{1 + \left(\frac{R_{in}}{R}\right)^{2m}}{1 - \left(\frac{R_{in}}{R}\right)^{2m}} \right) + \sum_{n=1}^{\infty}\sum_{m=1}^{\infty} \frac{c_{m,n}^2}{m^2 n^3} f_m(\gamma_n' R_{in}, \gamma_n' R)}, \qquad (5)$$

being C_1 a constant, which depends on the geometric characteristics of the sensor and the pipe, and γ_n' a constant that depends on the harmonic order n:

$$C_1 = \frac{2\pi^3 s_z^2 R}{H^3} \text{ and } \gamma_n' = \frac{2n\pi}{H}. \qquad (6)$$

Finally, the constant $c_{m,n}$ is given by the expression:

$$c_{m,n} = a_m(\theta_1, \theta_2) \sin\left(\frac{\gamma_n' s_z}{2}\right), \qquad \text{with } a_m(\theta_1, \theta_2) = \cos(m\theta_2) - \cos(m\theta_1), \qquad (7)$$

where, θ_1 and θ_2 denote the initial and final azimuthal coordinates of the emitter electrode, where it has been assumed that the azimuthal coordinates of the receiver electrode are, $-\theta_1$ and $-\theta_2$.

2.2 The relative conductance for hold up applications

When we have a two-phase flow mixture with average void fraction α inside the region of the pipe covered by the sensor and we denote by $G_\alpha^* = G_\alpha / (\sigma_{eff} \pi D)$, the non-dimensional conductance of the pipe full of the two-phase mixture, being σ_{eff} the effective electrical conductivity of the mixture. In addition, we denote by $G_{max}^* = G_{max}/(\sigma_w \pi D)$ to the non-dimensional sensor conductance when the pipe is full of the liquid phase with electrical conductivity σ_w. Then if we use a two-ring electrode sensor with conductance given by the expression [1], [2]:

$$G_{max}^* = \frac{G_{max}}{\sigma_w \pi D} = \frac{\pi^3}{8}\left(\frac{s_z}{H}\right)^2 \frac{1}{\sum_{n=0}^{\infty} \frac{b_n^2}{(2n+1)^3} \frac{I_0(\gamma_n R)}{I_1(\gamma_n R)}},$$
(8)

being the meaning of the symbols the same ones that in the previous section. Then the conductance ratio of the pipe full of the two-phase mixture and the pipe full of water is given by:

$$\frac{G_\alpha}{G_{max}} = \frac{G_\alpha^*}{G_{max}^*} \frac{\sigma_{eff}}{\sigma_w} = \frac{\sigma_{eff}}{\sigma_w},$$
(9)

Eqn (9) has been used by different author to measure the liquid holdup $\alpha_l = 1 - \alpha$. The problem is that to obtain α directly from the relative conductance value, this last value depends not only on the average void fraction but also on its distribution as will be discussed later.

2.3 The useful range of thicknesses of a conductance probe

As the film thickness becomes larger the variations ΔG^* in the response of the conductance-probe to variations $\Delta \delta$ of the film thickness becomes smaller, this means that the range of thicknesses that the probe can measure with good accuracy is limited. For this reason, Coney defined an error factor f_ε as follows [15]:

$$f_\varepsilon = \lim_{\Delta G^* \to 0} \frac{\Delta \delta/\delta}{\Delta G^*/G^*} = \frac{G^*}{\delta}\frac{d\delta}{dG^*}.$$
(10)

It is more convenient to express this factor in terms of the relative conductance with respect to the maximum conductance. The maximum conductance value is attained when the pipe is full of water, and the relative conductance G_{rel}^* is defined as the ratio of the conductance for a given annular flow and the value of the conductance for the pipe full of liquid:

$$G_{rel}^* = \frac{G^*}{G_{max}^*}.$$
(11)

Dividing the numerator and the denominator of eqn (10) by G_{max}^* and because of the maximum conductance is independent of δ, we obtain the following equation:

$$f_\varepsilon = \frac{G_{rel}^*}{\delta}\frac{d\delta}{dG_{rel}^*}.$$
(12)

From eqn (12) it is deduced that the error $\Delta \delta$ obtained in the measurement of the film thickness is related to error ΔG_{rel}^* obtained in the measurement of the relative conductance by the following approximate expression:

$$\frac{\Delta \delta}{\delta} \cong f_\varepsilon \frac{\Delta G_{rel}^*}{G_{rel}^*}.$$
(13)

One of the applications of the error factors f_ε suggested by Coney [15], is to compare the useful ranges of different conductance sensors.

3 COMPUTED RESULTS FOR DIFFERENTS CONDUCTANCE DESIGNS

3.1 Computed results with plate probes of parallel electrodes

Fig. 2 displays the results obtained for G^*_{rel} versus the water film thickness δ for two parallel plate electrodes when we change the width of the electrodes. The radius of the pipe was fixed at R = 25.4 mm, and the distance between the electrodes was also fixed at $2a = 2$ mm. It is observed at Fig. 2 that when the width of the electrodes become smaller maintaining the same length at $s_z = 2.88$ mm, then the relative conductance saturates for smaller film thicknesses i.e. it attains values which are closer to its maximum for smaller film thicknesses.

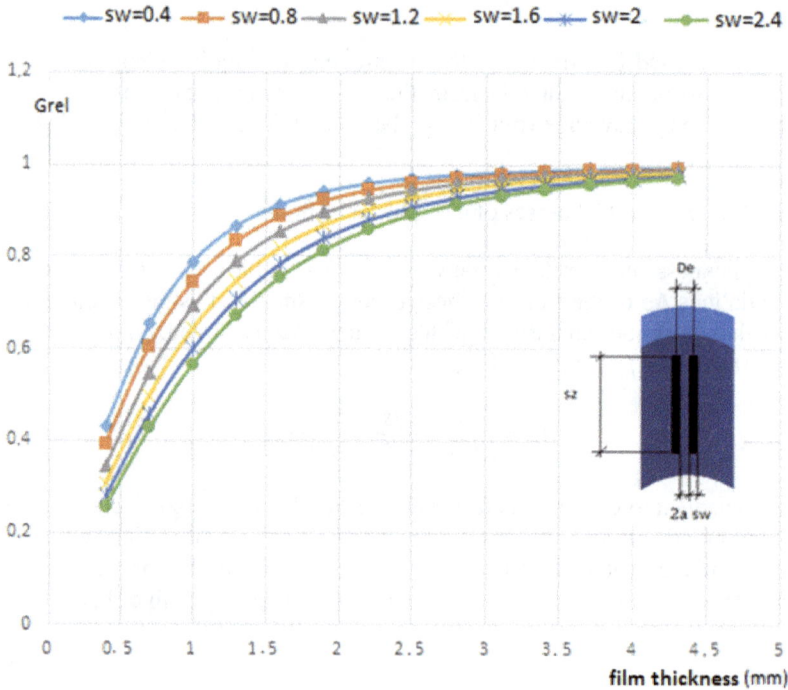

Figure 2: G^*_{rel} versus the film thickness δ (mm) for a sensor design with two parallel plate geometry and dimensions $s_z = 2.88$ mm, $a = 1$ mm, $R = 25.4$ mm, for different values of s_w in mm.

Therefore, when the conductance is close to its saturation, then small changes in the conductance value produces big changes in the film thickness, and the errors in the film thickness measurements tend to increase. For this same case, we have computed the error factor f_ε, by increasing the film thickness of the previous cases in $\Delta\delta = 0.05$ mm, and obtaining the corresponding increment in the film conductance ΔG^*_{rel}, and the applying eqn (13) to obtain the error factor values. The results are displayed in Fig. 3.

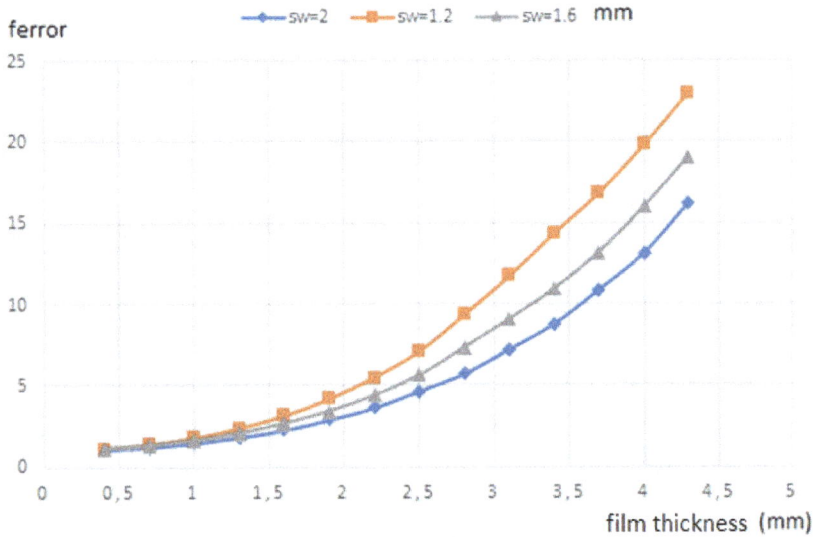

Figure 3: f_ε (error factor) versus the film thickness δ (mm) for a sensor design with two parallel plates geometry, and dimensions $s_z = 2.88$ mm, $a = 1$ mm, $R = 25.4$ mm, for different values of s_w in mm.

Fig. 3 tells us two things, the first is that when the film thickness increases the error factor also increases, this means that the conductance probe loss progressively its sensitivity when the film becomes thicker. Therefore, small errors in the conductance produce large errors in measuring the film thickness for large thicknesses of the conducting film. The solution to ameliorate this situation is to make larger the width of the electrodes. Increasing the size of the electrodes from 1.2 mm to 2 mm reduces the error factor at 1.9 mm from 4.16 to 2.91.

3.2 Computed results with plate probes in the flow direction

In two-phase flow tomography of annular flows, we are interested in measuring the height, shape, and frequency of the different types of waves, which are formed at the interface between the gas and liquid phases. These waves are mainly of two types: the first are the disturbance waves, which are prominent waves with a height that is approximately four times the thickness of the film and that are coherent in both the streamwise direction and circumferentially. The second class are the ripple waves which are not coherent in any direction and are much smaller in size. Also, although less frequents than the previous ones we can encounter the ephemeral waves, which have also large amplitudes and are no coherent. To measure the characteristics of these waves propagating in the streamwise direction we use conductance plate electrodes, as the ones displayed in Fig. 1(b), which should be as small as possible to capture the characteristic details of these waves. The problem is that if the waves become too large in amplitude and the size of the probe is two small the conductance probe can saturate. For this reason, it is convenient to be able to design the geometric characteristics of the probes (size of the electrodes and distance between them), for each specific application. Fig. 4 displays the relative conductance versus the film thickness δ in mm for plate conductance probes flush mounted in a pipe of 25.4 mm of radius.

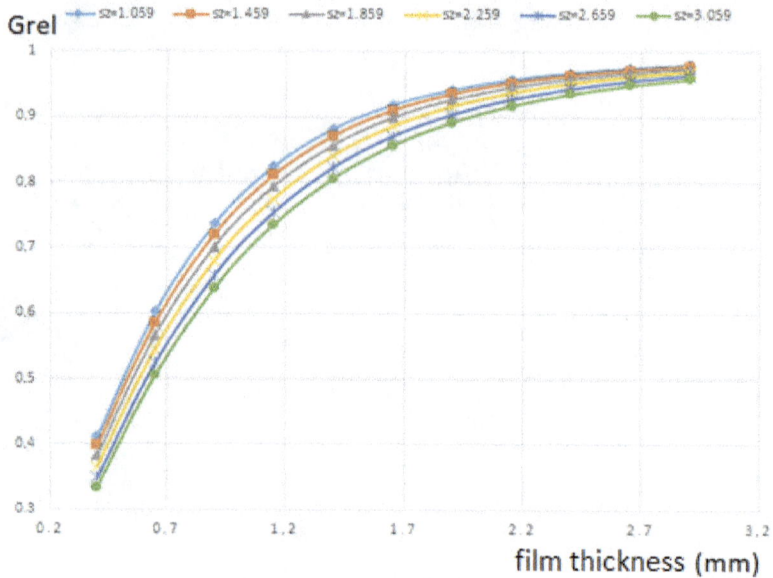

Figure 4: G^*_{rel} versus the film thickness δ (mm) for a sensor design with two plate electrodes along the flow direction with geometry, and dimensions s_w = 2.6586 mm, a = 1 mm, R = 25.4 mm, for different values of s_z in mm.

It is observed in Fig. 4 that the conductance probe attains earlier the saturation for smaller film thicknesses as the electrode length s_z in the streamwise direction becomes smaller. The considered electrode geometries with distances $D_e = 2a + s_z$ between the centre of the electrodes ranging from 3 mm to 5 mm, provide good estimation of the film thickness in the range from 0.2 to 1.7 mm. Although for values of the thickness above 1.5 mm the errors become larger. Fig. 5 displays the error factors for s_z=1.059 mm, s_z = 1.859 mm, and s_z = 3.059 mm, we notice that the error factors increase with the thickness of the film. For the conductance sensors with the smaller electrodes about 1 mm, we have acceptable error factors of 4 below a film thickness of 1.5 mm. However, for the sensor with s_z = 3.059 mm the factor of error is below 4.2 for film thickness smaller than 2 mm.

3.3 Application of the conductance probe to holdup determination

The application of two-ring conductance probes to holdup applications allows to obtain the liquid fraction or the void fraction values directly from the relative conductance of the gas liquid mixture to that of the pipe full of the conducting liquid (water). If the void fraction distribution in the mixture is homogeneous then as found by Tshochatzidis [2], the Maxwell equation for the effective conductivity provides the best prediction of the liquid fraction from the relative conductance [1], [2]. However, if the mixture becomes non-homogeneous then Muñoz-Cobo et al. [1] found that the expression that best fitted the experimental data of Yang et al. [12] was the effective medium theory (EMT) in the self-consistent approximation. However, this approximation fails for values of the liquid fraction below 0.86, as observed in Fig. 6. To try to improve these results we have used the percolation theory in continuous

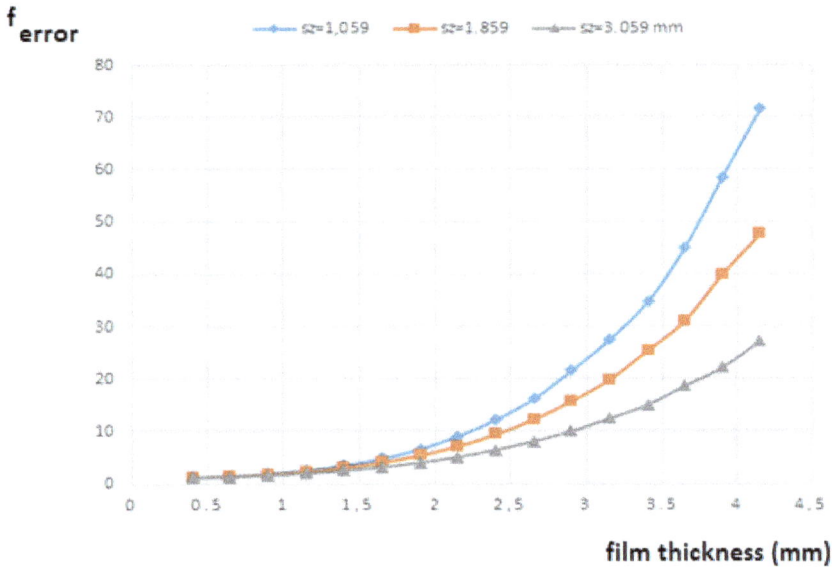

(a)

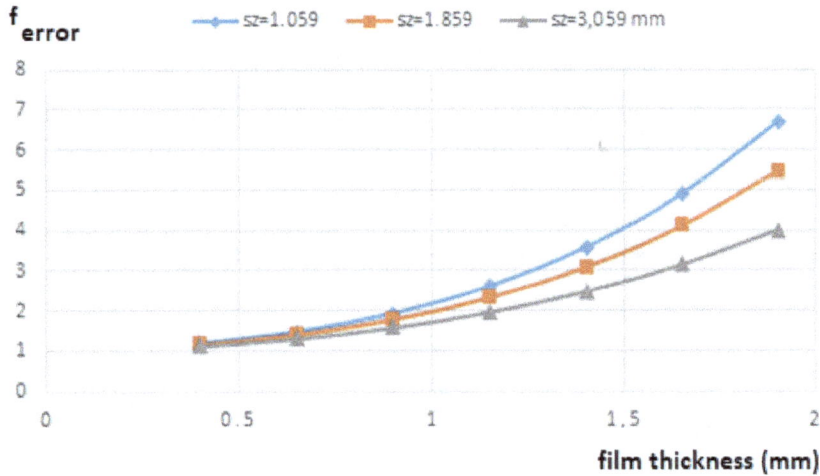

(b)

Figure 5: Error factor f_ε versus the film thickness δ (mm) for a sensor design with two plate electrodes along the flow direction with geometry, and dimensions $s_w = 2.6586$ mm, a = 1 mm, R = 25.4 mm, for different values of s_z in mm. (a) For film thickness up to 4.5 mm; and (b) For film thickness up to 2 mm.

media, which provides much better predictions than the previous ones for non-homogeneous bubbly flows as observed in Fig. 6. This theory is especially applicable when the electrical conductivities of both phases of the two-phase mixture differs significantly as happens in the present case where the conductivity of the gas phase is negligible [18], [19].

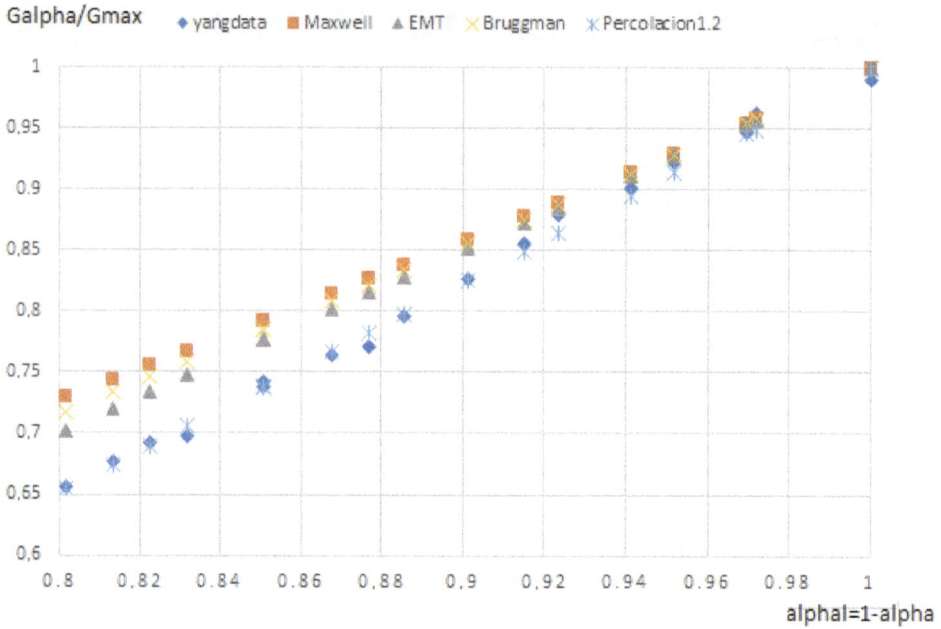

Figure 6:   Relative conductance of the two-phase mixture versus the liquid fraction $\alpha_l = 1 - \alpha$, for different models Maxwell (squares), self-consistent EMT (triangles), Bruggman (x), percolation (blue x with a vertical line) and Yang et al. [12] data (rhomboids).

The percolation theory used to predict the effective conductivity of the two-phase mixture gives the following expression for the effective conductivity [18], [19]:

$$\sigma_{eff} = \sigma_w \left(\frac{\alpha_l - \alpha_c}{1 - \alpha_c}\right)^{\mu} \; for \; \alpha_l > \alpha_c \; and \; \sigma_{eff} = 0 \; for \; \alpha_l < \alpha_c, \qquad (14)$$

where α_c is the percolation threshold for the liquid fraction of the conducting phase that is equal to 1/3, α_l is the volume fraction of the conducting phase, finally μ is the critical exponent. The calculations of the effective conductivity using the percolation model have been performed with a percolation threshold $\alpha_l = 1/3$, and a critical exponent $\mu = 1.2$.

Finally Fig. 7 displays the relative conductance versus the liquid fraction computed with the percolation model for critical exponents of 1.15, 1.2 and 1.25 we see that the three values match the experimental data, but the critical exponent of 1.25 gives results which are a little lower than the experimental ones for the lower liquid fractions.

4 CONCLUSIONS

In this paper we have discussed how to obtain the best design of the conductance probes, which are used for two-phase tomography mainly of annular flows and hold up applications as determining the average liquid fraction or the average void faction inside a pipe containing a two-phase flow mixture. We have performed the calculation of the useful range of the conductance probes by means of the error factor that gives as expressed by eqn (13), how is

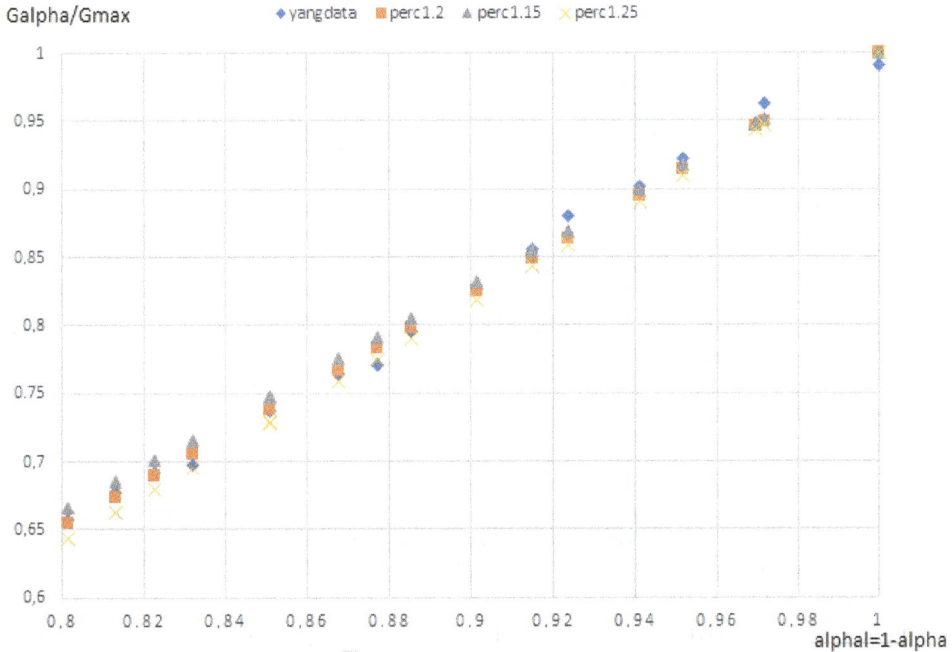

Figure 7: Relative conductance versus the liquid fraction for Yang experiments [12], and the percolation model computed with the critical exponents of 1.15, 1.2 and 1.25.

transformed the relative error in the conductance to the relative error in the film thickness. Because the conductance depends on the conductivity and the conductivity changes with the temperature, these error factors give us if the conductivity is maintained within a given error for instance 4%, the error that we can have when measuring the film thickness. For instance, if the error factor for a conductance sensor design is 4, and the conductivity is maintained constant within a 4%, then we will have an error in δ of 16%. If we want a tolerance for δ smaller than 10%, then with the curve of Fig. 5, we shall use a sensor design with $s_z = 3$ mm, with this height of the electrodes and the geometry given in the caption of Fig. 5, we are limited to measure film thickness smaller than 1.4 mm, to have a tolerance smaller than 10%.

Another important issue studied in this paper is the prediction of the liquid holdup or the void fraction when the bubbly flow is not homogeneous, we have seen that in this case the best prediction of Yang et al. [12] data is obtained with the percolation theory of 3D continuous systems, in this case the threshold and the critical exponents are computed by Monte-Carlo although many authors [18] recommend considering the critical exponent μ as a fitting parameter. The calculations of Fig. 6 were performed with a critical exponent value of $\mu = 1.2$.

ACKNOWLEDGEMENT

The authors of this paper are indebted to the financial support received from Spain Ministry of Science and Technology contract number ENE2016-79489-C2-1-P (EXMOTRANSIN), of the Plan Nacional de I + D.

REFERENCES

[1] Muñoz-Cobo, J.L., Rivera, Y., Berna, C. & Escrivá, A., Analysis of conductance probes for two-phase flow and holdup applications. *Sensors*, **20**, p. 7042. DOI: 10.3390/s20247, 2020.

[2] Tsochatzidis, N.A., Karapantsios, T.D., Kostoglou, M.V. & Karabelas, A.J., A conductance probe for measuring liquid fraction in pipes and packed beds. *International Journal of Multiphase Flow*, **18**(5), pp. 653–667, 1992.

[3] Collier, J.G. & Thome, J.R., *Convective Boiling and Condensation, Oxford Engineering Science Series*, Editorial Clarendon Press, pp. 1–16, 1994.

[4] Belt, R.J., Vant't Westende, J.M.C., Prasser, H.M. & Portela, L.M., Time and spatially resolved measurements of interfacial waves in vertical annular flow. *International Journal of Multiphase Flow*, **36**, pp. 570–587, 2010.

[5] Zhao, Y., Markides, C. & Hewitt, G.F., Disturbance wave development in two-phase gas-liquid upward vertical annular flow. *International Journal of Multiphase Flow*, **55**, pp. 111–129, 2013.

[6] Rivera, Y., Muñoz-Cobo, J.L., Cuadros, J.L., Berna, C. & Escrivá, A., Experimental study of the effects produced by the changes of the liquid and gas superficial velocities and the surface tension on the interfacial waves and the film thickness in annular concurrent upward vertical flows. *Experimental Thermal and Fluid Science*, **120**, p. 110224, 2021.

[7] Berna, C., Escrivá, A., Muñoz-Cobo, J.L. & Herranz, L., Review of droplet entrainment in annular flow: Interfacial waves and onset of entrainment. *Progress in Nuclear Energy*, **74**, pp. 14–43, 2014.

[8] Lopes, J.C.B. & Dukler, A.E., Dopler entrainment in vertical annular flow and its contribution to momentum transfer. *AICHE Journal*, **32**(9), pp. 1500–1515, 1986.

[9] Munoz-Cobo, J.L., Chiva, S., Mendes, S., Monrós, G., Escrivá, A. & Cuadros, J.L., Development of conductivity sensors for multi-phase flow local measurements at the Polytechnic University of Valencia (UPV) and University Jaume I of Castellon (UJI). *Sensors*, **17**, p. 1077, 2017. DOI: 10.3390/s17051077.

[10] Manera, A., Ozar, B., Paranjape, S., Ishii, M. & Prasser, H., Comparison between wire-mesh sensors and conductive needle-probes for measurements of two-phase flow parameter. *Nuclear Engineering and Design*, **239**, pp. 1718–1724, 2009.

[11] Prasser, H.M., Boettger, A. & Zschau, J., A new electrode-mesh tomography for gas-liquid flows. *Flow Measurement and Instrumentation*, **9**, pp. 111–119, 1998.

[12] Yang, H.C., Kim, D.K. & Kim, M.H., Void fraction measurement using impedance method. *Flow Measurement and Instrumentation*, **14**, pp. 151–160, 2003.

[13] Andreussi, P., Donfrancesco, A.D. & Messia, M., An impedance method for the measurement of liquid hold-up in two-phase flow. *International Journal of Multiphase Flow*, **14**, pp. 777–785, 1988.

[14] Jones, O.C. & Zuber, N., The interrelation between void fraction fluctuations and flow patterns in two-phase flow. *International Journal of Multi-Phase Flow*, **2**, pp. 273–306, 1975.

[15] Coney, M.W.E., The theory and application of conductance probes for the measurement of liquid film thickness in two phase flow. *Journal of Physics E: Scientific Instruments*, **6**, pp. 903–910, 1973.

[16] Zadrazil, I., Matar, O.K. & Markides, C.N., An experimental characterization of downwards gas–liquid annular flow. *International Journal of Multiphase Flow*, **60**, pp. 87–102, 2014.

[17] Fossa, M., Design, and performance of a conductance probe for measuring the liquid fraction in two-phase gas-liquid flows. *Flow Measurement and Instrumentation*, **9**, pp. 103–109, 1998.

[18] Kovacik, J., Electrical conductivity of two-phase composite materials. *Script Materialia*, **39**(2), pp. 153–157, 1998.

[19] Stauffer, D. & Aharony, A., *Introduction to Percolation Theory*, 2nd edn, Taylor and Francis: London, 1992.

SIMULATION OF FLUID'S AERATION: IMPLEMENTATION OF A NUMERICAL MODEL IN AN OPEN SOURCE ENVIRONMENT

MARCO NICOLA MASTRONE & FRANCO CONCLI
Free University of Bolzano, Italy

ABSTRACT

Multiphase flows often involve additional complex physical phenomena to be considered, for example cavitation and aeration. These can arise in several applications as gear pumps, propellers, and gearboxes. In order to properly design such hydraulic systems, these aspects should be carefully considered in the design phase. In fact, these phenomena can lead to lower efficiency, excessive noise, and a rapid degradation of the mechanical system. While for cavitation analytical and numerical models are already available, aeration, which refers to the entrainment of air in another fluid in form of bubbles and/or foam, is more difficult to be quantified. Even experimentally niche equipment is required. However, thanks to recent developments in computer science, it is possible to approach these complex multiphase flows exploiting simulations' tools. In this paper, a numerical approach based on Computational Fluid Dynamics (CFD) that includes the aeration phenomenon is presented. A new solver that accounts for the air entrapment in the fluid was implemented in the open source environment OpenFOAM®. The solver was applied on a test case to verify its behaviour and compared to a standard multiphase solver. The qualitative analysis suggests that the solver is capable of considering the amount of air entrapped in the fluid, thus opening a new way in the study of aeration.
Keywords: aeration, CFD, multiphase modelling, OpenFOAM.

1 INTRODUCTION

Thanks to continuous technological developments in computer science, simulation software packages have gained their consideration not only among academic researchers, but also in industries. The advantages introduced by virtual prototyping, namely the reduction of time and costs with respect to real prototypes, have encouraged engineers to exploit these new opportunities. Complex multiphase problems can be solved numerically exploiting dedicated software where proper boundary conditions are assigned to each part of the computational domain. The study of multiphase problems is of great interest for real engineering applications as lubrication mechanisms, design of hydraulic systems, and in general every condition where more fluids are present simultaneously.

Standard multiphase solvers are not capable of considering additional physical phenomena, as cavitation and aeration, which play a determinant role in several operating conditions and must be carefully considered for a proper design of a system to avoid efficiency reduction, excessive noise, and the rapid degradation of the mechanical system. In fact, in order to include also possible phase changes (cavitation) and air entrainment in another fluid in form of bubbles and/or foam (aeration), the conservation equations of fluid dynamics must be modified by adding a source term that accounts for the new phenomena. While works dealing with standard multiphase solvers [1]–[13] and the simulation of cavitation in the design of mechanical components are already available [14]–[19], the aeration has been poorly touched from a numerical point of view. The development of a numerical solver that includes the air trapping in another fluid can lead to significant advancements in the correct modelling of real physical problems. Indeed, in hydraulic systems, the presence of air can cause the premature degradation of the lubricant and the increase of wear of mechanical components. Moreover, aeration has a severe impact on the

WIT Transactions on Engineering Sciences, Vol 132, © 2021 WIT Press
www.witpress.com, ISSN 1743-3533 (on-line)
doi:10.2495/MPF210031

heat transfer capabilities of the lubricant, thus reducing the global efficiency of the system and the heat dissipation capability, which, in turn, is responsible for overheating and possible premature failures.

In order to quantify the level of aeration in various operating conditions, niche and expensive equipment is required [20]–[23]. Hence, CFD offers the opportunity to investigate this complex phenomenon numerically. A numerical approach to calculate the air entrainment in a fluid was proposed by Cerne et al. [24], who established an interface tracking algorithm based on a two-fluid model formulation [25]. Yan and Che [26] extended the previous model for three fluid phases. A combination of Eulerian approach and Volume of Fluid (VOF) was tested by Wardle and Weller [27]. Ma et al. [28] reformulated the source term proposed by Sene [29] by including the turbulence intensity and tested their model on an hydraulic jump case [30], a plunging jet [31] and flow around a ship [32]. Several experiments related to plunging jet [33]–[36] have been exploited to study aeration, since, due to the energy dissipation, air bubbles are generated at the water free surface. In this work, a solver that considers aeration is implemented in the open source software OpenFOAM® and applied to two test cases. The promising results suggest that numerical tools can offer an opportunity to investigate conditions where experimental data might be difficult to obtain due to the complexity of the system and the required equipment.

2 MATERIALS AND METHODS

2.1 Mathematical description

CFD codes are based on the solution of three governing equations: mass, momentum, and energy conservation. In this study, the problem was modelled as isothermal. Therefore, the energy equation was not included in the calculation. In this way, the solution is limited to the mass and momentum equations which can be written as:

$$\frac{\partial \rho \langle u_i \rangle}{\partial x_i} = 0. \tag{1}$$

$$\frac{\partial (\rho \langle u_i \rangle)}{\partial t} + \frac{\partial (\rho \langle u_i \rangle \langle u_i \rangle)}{\partial x_j} = -\frac{\partial \langle p \rangle}{\partial x_i} + \frac{\partial}{\partial x_j}\left[\mu\left(\frac{\partial \langle u_i \rangle}{\partial x_j} + \frac{\partial \langle u_j \rangle}{\partial x_i}\right)\right] - \frac{\partial \tau_{ij}}{\partial x_j}. \tag{2}$$

The term τ_{ij} is the Reynolds term and is expressed exploiting the eddy viscosity (μ_t) hypothesis:

$$-\tau_{ij} = \mu_t\left(\frac{\partial \langle u_i \rangle}{\partial x_j} + \frac{\partial \langle u_j \rangle}{\partial x_i}\right) - \frac{2}{3}\rho\delta_{ij}k. \tag{3}$$

$$\mu_t = C_\mu \rho \frac{k^2}{\varepsilon}. \tag{4}$$

Turbulence is modeled according the RNG $k - \varepsilon$ model:

$$\frac{\partial}{\partial t}(\rho k) + \frac{\partial}{\partial x_i}(\rho k u_i) = \frac{\partial}{\partial x_j}\left[\alpha_k \mu_{eff}\frac{\partial k}{\partial x_j}\right] + G_k + G_b - \rho\varepsilon - Y_M + S_k. \tag{5}$$

$$\frac{\partial}{\partial t}(\rho\varepsilon) + \frac{\partial}{\partial x_i}(\rho\varepsilon u_i) = \frac{\partial}{\partial x_j}\left[\alpha_\varepsilon \mu_{eff}\frac{\partial k}{\partial x_j}\right] + \frac{C_{1\varepsilon}\varepsilon}{k}(G_k + C_{3\varepsilon}G_b) - \frac{C_{2\varepsilon}\rho\varepsilon^2}{k} - R_\varepsilon + S_\varepsilon. \tag{6}$$

In these equations k is the turbulence kinetic energy, ε is the dissipation rate, G_k and G_b represent the turbulence kinetic energy that originates from velocity gradients and from buoyancy effects respectively, Y_M stands for the fluctuating dilation in compressible turbulence. $C_{1\varepsilon}$, $C_{2\varepsilon}$ and $C_{3\varepsilon}$ are constants. α_k and α_ε are the turbulent Prandtl numbers. S_k and S_ε are source terms. R_ε is a term deriving from the renormalization group theory that characterizes the RNG model.

These equations are valid only in simulations involving one phase. In order to model multiphase problems numerically, an additional balance equation to consider the presence of two or more phases must be added to the previous equation. By exploiting the VOF model [37], which is based on the definition of the scalar quantity volumetric fraction representing the percentage of one fluid in every cell of the domain, the multiphase problem can be solved. The equation of the volumetric fraction can be expressed as follows:

$$\frac{\partial \alpha}{\partial t} + \nabla(\alpha \boldsymbol{u}) = 0. \tag{7}$$

The properties Θ of the different fluids (such as density and viscosity) are taken to define the properties of an equivalent fluid as follows:

$$\Theta = \Theta \alpha + \Theta(1 - \alpha), \tag{8}$$

where Θ represents the generic property of each fluid.

The MULES (Multidimensional Universal Limiter with Explicit Solution) [38] correction can be added in the solver algorithm in order to obtain a more stable and bounded solution of the volumetric fraction field. This is accomplished by adding a dummy velocity field ($\boldsymbol{u}_c$) in the conservation equation of the volumetric fraction:

$$\frac{\partial \alpha}{\partial t} + \nabla(\alpha \boldsymbol{u}) + \nabla(\boldsymbol{u}_c \alpha(1 - \alpha)) = 0. \tag{9}$$

An additional source term (S_g) must be added to the equation to account for additional phenomena as cavitation and aeration:

$$\frac{\partial \alpha}{\partial t} + \nabla(\alpha \boldsymbol{u}) + \nabla(\boldsymbol{u}_c \alpha(1 - \alpha))) = S_g. \tag{10}$$

To calculate the source term, a mathematical model must be introduced. The most used ones for describing cavitation are those by Kunz [39], Merkle [40] and Saurer [41] while aeration can be described using the model by Hirt [42]. In the latter, air is entrained into the water when turbulent energy per unit volume is larger than the stabilizing force of gravity and surface tension per unit volume. The expression for this source term is given as:

$$S_g = C_{air} A_S \sqrt{2 \frac{P_t - P_d}{\rho}}, \tag{11}$$

where S_g is the volume of air per unit of time, C_{air} is a calibration parameter, A_S the free surface area at each cell, and P_t (turbulent forces) and P_d (stabilizing forces) are given by:

$$\boldsymbol{P_t} = \rho k, \tag{12}$$

$$P_d = \rho g_n L_T + \frac{\sigma}{L_T},$$ (13)

where g_n is the normal component of the gravity, σ is the surface tension, and L_T is the turbulence characteristic length scale, expressed as:

$$L_T = C_\mu \cdot \sqrt{\frac{3}{2}} \frac{k^{\frac{3}{2}}}{\varepsilon},$$ (14)

where $C_\mu = 0.085$ for the RNG $k - \varepsilon$ turbulence model. An air volume is added to an element of the computational domain when the stabilizing forces P_d are exceeded by the perturbing forces P_t, according to eqn (11). C_{air} is the coefficient that appears in the equation of the source term that should be calibrated. In this case it is set to 0.5. Hirt used this value to validate their model. For most of the cases this value has proven to be reasonable. This coefficient indicates the percentage of the free surface area on which air entrapment occurs.

2.2 Solver settings

The PIMPLE (merged PISO-SIMPLE) algorithm was used in the simulations. A convergence criterion of 1e-6 was imposed to all field's variables. The Preconditioned Conjugate Gradient (PCG) solver was used for the pressure solver, while the Gauss-Seidel smooth solver was used for the velocity. The Courant number was limited to 0.5 to ensure the stability of the simulations. The first-order implicit Euler scheme was used for time-stepping. A Total Variation Diminishing (TVD) scheme using the vanLeer limiter was adopted for the convection of the volumetric fraction. The convective fluxes in the turbulence equations were discretized using first order schemes.

3 RESULTS

Two test cases were considered to study the behaviour of the solver: a vertical plunging jet and a free fall of a droplet. Each of these can be considered as a benchmark for the solver since they can involve some air entrapment at free surface. The introduction of the new solver is expected to improve the physical representation of the real problem due to its capability to consider the air entrainment at free surface.

3.1 Vertical plunging jet

In most cases of aeration, air entrapment occurs at a free surface discontinuity where large velocities gradients come across. For this reason, a test case consisting of a jet penetrating into a water pool is considered to test the behaviour of the solver. In Fig. 1 the computational domain is shown. The geometry was taken from Hirt [42] who used it in his validation tests; the problem was modelled as 2D (a box with dimensions of 0.5×0.5 m^2) and the domain was meshed with pure quadrilateral elements (about 25k cells; the geometry was properly partitioned to increase the elements' density in the contact region between the jet and the free surface).

An inlet velocity of 0.4 m/s is applied at the jet patch, where turbulent quantities are also imposed, while the pressure must be calculated. Wall functions are applied to the other walls of the domain regarding the turbulent quantities, while the velocity is set to zero.

In Fig. 2 the comparison between the standard multiphase solver of OpenFOAM® and the new one that considers aeration is shown in terms of volumetric fraction field. Two time steps are reported (initial transitory and stabilized filed at regime).

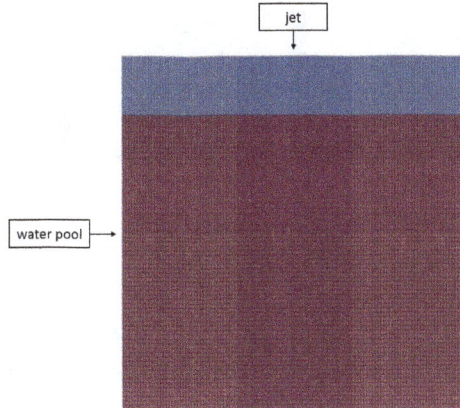

Figure 1: Computational domain. A finer mesh is implemented at the air-water interface.

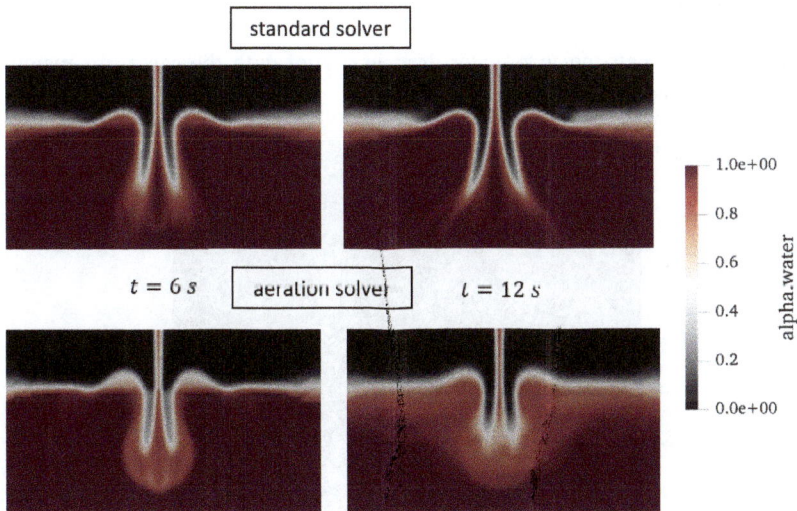

Figure 2: Standard (top row) and aeration (bottom row) volumetric field contour.

It can be observed that the standard multiphase solver promotes a sharp interface between water and air, while the new solver shows a smoother transition at the interface showing that a non-negligible amount of air is entrapped in the water at free surface. This means that, at least qualitatively, the physics of the problem is captured from the solver. Moreover, the contours clearly indicate that, in the case of the standard multiphase solver, the jet substantially enters the steady sump, while the new solver slows down the main flow that tends to diffuse radially due to the presence of foaming effects. Indeed, the foam tends to slow down the water jet penetration forcing it to move laterally.

The effects of the source term at four different time steps are reported in Fig. 3. It can be noticed that it is active on a large region of the free surface, thus indicating that aeration is occurring.

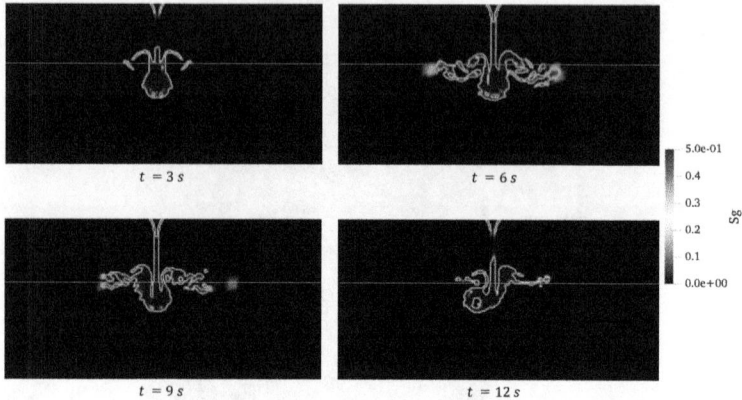

Figure 3: Source term contour at four different time steps.

3.2 Droplet free fall

A second test case consisting in a water drop in a pool with the same dimensions of the previous case is considered (Fig. 4).

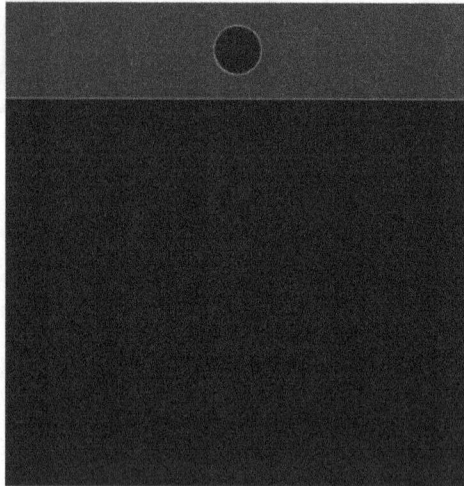

Figure 4: Domain consisting of a water droplet and a water pool.

In this simulation only the gravity is responsible for the droplet movement. In Fig. 5, the volumetric fraction and the effects of the source term are represented.

In this case, aeration seems to be minor with respect to the previous case. In fact, the source term is active on a very small region of the domain and the air trapping is not so evident. This suggests that while in the plunging jet case, the imposition of a flow rate to the fluid forced the aeration phenomenon to take place, in the droplet case the only effect of gravity is not sufficient to cause evident foaming effects at free surface. Therefore, the solver seems to be able to add air entrainment only when it actually occurs ($P_t > P_d$).

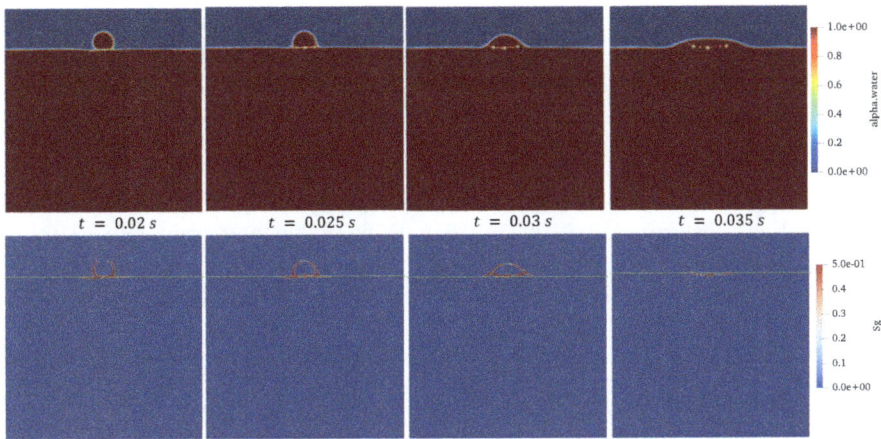

Figure 5: Volumetric fraction and source term contours.

Even if these are relatively simple cases, the solver behaves as expected, at least conceptually, suggesting its being a possible tool for the investigation of the aeration phenomenon. The analysis of the simulations indicates that, at least qualitatively, the solver implemented according to the Hirt model is capable of considering the air entrainment, and, possibly, can be extended to study other hydraulic systems providing at least some design information that would be complex to obtain experimentally.

4 CONCLUSIONS

The introduction of CFD has opened new ways in the approach of real engineering problems. Indeed, the benefits of virtual prototypes in terms of cost and time reduction have encouraged designers to exploit simulation tools to support experimental findings. Multiphase problems represent a challenging task in simulations due to the complexity of the required models. In this work, a solver that considers the air entrainment based on the Hirt model is implemented in the open source environment OpenFOAM®. The solver was tested on a plunging jet and the results compared with the native multiphase solver of the software used. The results show that, at least qualitatively, the model adds some air at free surface. This solver may be possibly used to analyse operating conditions where the adoption of a standard multiphase solver is not sufficient to capture the generation of aeration. In future, it is planned to apply the solver to different test cases (e.g. to simulate the aeration inside gearboxes).

REFERENCES

[1]   Mastrone, M.N., Hartono, E.A., Chernoray, V. & Concli, F., Oil distribution and churning losses of gearboxes: Experimental and numerical analysis. *Tribology International*, **151**, p. 106496, 2020.

[2]   Burberi, E., Fondelli, T., Andreini, A., Facchini, B. & Cipolla, L., CFD simulations of a meshing gear pair. *Proceedings of the ASME Turbo Expo*, vol. 5A-2016, American Society of Mechanical Engineers (ASME), 2016.

[3]   Concli, F. & Gorla, C., CFD simulation of power losses and lubricant flows in gearboxes. *American Gear Manufacturers Association Fall Technical Meeting*, 2017.

[4]   Concli, F. & Gorla, C., A CFD analysis of the oil squeezing power losses of a gear pair. *International Journal of Computational Methods & Experimental Measurements*, **2**, pp. 157–167, 2014.

[5]   Concli, F., Maccioni, L. & Gorla, C., Lubrication of gearboxes: CFD analysis of a cycloidal gear set. *WIT Transactions on Engineering Sciences*, vol. 123, ed. S. Hernandez, WIT Press: Southampton and Boston, pp. 101–112, 2019.

[6]   Concli, F., Schaefer, C.T. & Bohnert, C., Innovative strategies for bearing lubrication simulations. *Preprints*, 2019100218, 2019. DOI: 10.20944/preprints201910.0218.v1.

[7]   Frosina, E., Senatore, A., Buono, D., Manganelli, M.U. & Olivetti, M., A tridimensional CFD analysis of the oil pump of an high performance motorbike engine. *Energy Procedia*, **45**, pp. 938–948, 2014.

[8]   Ferrari, C. & Marani, P., Study of air inclusion in lubrication system of CVT gearbox transmission with biphasic CFD simulation. *Symposium on Fluid Power and Motion Control*, Bath, UK, 2016. DOI: 10.1115/FPMC2016-1767.

[9]   Santra, T.S., Raju, K., Deshmukh, R., Gopinathan, N., Paradarami, U. & Agrawal, A., Prediction of oil flow inside tractor transmission for splash type lubrication. *SAE Technical Papers*, 2019-01-09, 2019.

[10]   Huang, S., Wei, Y., Huo, C. & Kang, W., Numerical simulation and performance prediction of centrifugal pump's full flow field based on OpenFOAM. *Processes*, 7(9), 605, 2019.

[11]   Concli, F., Low-loss gears precision planetary gearboxes: Reduction of the load dependent power losses and efficiency estimation through a hybrid analytical-numerical optimization tool. Hochleistungs- und Präzisions-Planetengetriebe: Effizienzschätzung und Reduzierun. *Forsch. im Ingenieurwesen/Engineering Res.*, **81**, pp. 395–407, 2017.

[12]   Petit, O. & Nilsson, H., Numerical investigations of unsteady flow in a centrifugal pump with a vaned diffuser. *International Journal of Rotating Machinery*, **2013**, Article ID 961580, 2013. DOI: 10.1155/2013/961580.

[13]   Concli, F. & Gorla, C., Influence of lubricant temperature, lubricant level and rotational speed on the churning power loss in an industrial planetary speed reducer: Computational and experimental study. *International Journal of Computational Methods & Experimental Measurements*, **1**, pp. 353–366, 2013.

[14]   Močilan, M., Husár, Š., Labaj, J. & Žmindák, M., Non-stationary CFD simulation of a gear pump. *Procedia Engineering*, **177**, pp. 532–539, 2017.

[15]   Gao, G., Yin, Z., Jiang, D. & Zhang, X., Numerical analysis of plain journal bearing under hydrodynamic lubrication by water. *Tribololy International*, **75**, pp. 31–38, 2014.

[16]   Sawicki, J. & Rao, T., Cavitation effects on the stability of a submerged journal bearing. *International Journal of Rotating Machinery*, **10**, pp. 227–232, 2004.

[17]   Concli, F., Pressure distribution in small hydrodynamic journal bearings considering cavitation: A numerical approach based on the open- source CFD code OpenFOAM®. *Lubrication Science*, **28**, pp. 329–347, 2016.

[18]   Concli, F. & Gorla, C., Numerical modeling of the power losses in geared transmissions: Windage, churning and cavitation simulations with a new integrated approach that drastically reduces the computational effort. *Tribololy International*, **103**, pp. 58–68, 2016.

[19]   Riedel, M., Schmidt, M. & Stücke, P., Numerical investigation of cavitation flow in journal bearing geometry. *EPJ Web of Conferences*, **45**, pp. 1–4, 2013.

[20] Borges, J.E., Pereira, N.H., Matos, J. & Frizell, K.H., Performance of a combined three-hole conductivity probe for void fraction and velocity measurement in air–water flows. *Experiments in Fluids*, **48**, pp. 17–31, 2010.

[21] Leandro, J., Bung, D.B. & Carvalho, R., Measuring void fraction and velocity fields of a stepped spillway for skimming flow using non-intrusive methods. *Experiments & Fluids*, **55**, pp. 1–17, 2014.

[22] Leprince, G., Changenet, C., Ville, F., Velex, P., Dufau, C. & Jarnias, F., Influence of aerated lubricants on gear churning losses – An engineering model. *Tribology Transactions*, **54**, pp. 929–938, 2011.

[23] Neurouth, A., Changenet, C., Ville, F., Octrue, M. & Tinguy, E., Experimental investigations to use splash lubrication for high-speed gears. *Journal of Tribology*, **139**(6), 061104, 2017.

[24] Cerne, G., Peterlin, S. & Tiselj, I., Coupling of the interface tracking and the two-fluid models for the simulation of incompressible two-phase flow. *Journal of Computational Physics*, **171**, pp. 776–804, 2001.

[25] Drew, D. & Passman, S., *Theory of Multicomponents Fluids*, Springer: New York, 1998.

[26] Yan, K. & Che, D., A coupled model for simulation of the gas–liquid two-phase flow with complex flow patterns. *International Journal of Multiphase Flow*, **36**, pp. 338–348, 2010.

[27] Wardle, K.E. & Weller, H.G., Hybrid multiphase CFD solver for coupled dispersed/segregated flows in liquid-liquid extraction. *International Journal of Chemical Engineering*, **2013**, Article ID 128936, 2013.

[28] Ma, J., Oberai, A., Drew, D., Lahey, R. & Moraga, F., A quantitative sub-grid air entrainment model for bubbly flows – plunging jets. *Computers & Fluids*, **39**, pp. 77–86, 2010.

[29] Sene, K.J., Air entrainment by plunging jets. *Chemical Engineering Science*, **43**, pp. 2615–2623, 1988.

[30] Ma, J., Oberai, A., Drew, D., Lahey, R. & Moraga, F., Modeling air entrainment and transport in a hydraulic jump using two-fluid RANS and DES turbulence models. *Heat & Mass Transfer*, **47**, pp. 911–919, 2011.

[31] Ma, J., Oberai, A., Drew, D., Lahey, R. & Moraga, F., A comprehensive sub-grid air entrainment model for RaNS modeling of free-surface bubbly flows. *Journal of Computational Multiphase Flows*, **3**, pp. 41–56, 2011.

[32] Ma, J., Oberai, A., Drew, D., Lahey, R. & Moraga, F., Two-fluid modeling of bubbly flows around surface ships using a phenomenological subgrid air entrainment model. *Computaters & Fluids*, **52**, pp. 50–57, 2011.

[33] Bayler, A. & Emiroglu, M.E., An experimental study of air entrainment and oxygen transfer at a water jet from a nozzle with air holes. *Water Environment Research*, **76**, pp. 231–237, 2014.

[34] Bin, A.K., Gas entrainment by plunging liquid jets. *Chemical Engineering Science*, **48**, pp. 3585–3630, 1993.

[35] Chanson, H., Aoki, S. & Hoque, A., Physical modelling and similitude of air bubble entrainment at vertical circular plunging jets. *Chemical Engineering Science*, **59**, pp. 747–754, 2004.

[36] Kiger, K.T. & Duncan, J.H., Air-entrainment mechanisms in plunging jets and breaking waves. *Annual Review of Fluid Mechanics*, **44**, pp. 563–596, 2012.

[37] Hirt, C.W. & Nichols, B.D., Volume of fluid (VOF) method for the dynamics of free boundaries. *Journal of Computational Physics*, **39**, pp. 201–225, 1981.

[38] Rusche, H., *Computational Fluid Dynamics of Dispersed Two-Phase Flows at High Phase Fractions*, Imperial College of Science, Technology and Medicine: London, 2002.

[39] Kunz, R.F. et al., Preconditioned Navier-Stokes method for two-phase flows with application to cavitation prediction. *Computers & Fluids*, **29**(8), pp. 849–875, 2000.

[40] Merkle, C.L., Feng, J. & Buelow, P.E.O., Computational modeling of the dynamics of sheet cavitation. *3rd International Symposium on Cavitation*, Grenoble, France, pp. 47–54, 1998.

[41] Saurer, J., Instationären kaviterende Sträömung - Ein neues Modell, basierend auf Front Capturing (VoF) and Blasendynamik, Universität Karlsruhe, 2000.

[42] Hirt, C.W., Modeling Turbulent Entrainment of Air at a Free Surface, Flow Science Report 01-12, Flow Science, Inc., 2003. DOI: 10.1061/40737(2004)187.

EXPERIMENTAL INVESTIGATION OF THE TRANSITION ZONE OF AIR–STEAM MIXTURE JETS INTO STAGNANT WATER

YAISEL CÓRDOVA[1,2], DAVID BLANCO[1], CÉSAR BERNA[1], JOSÉ LUIS MUÑOZ-COBO[1],
ALBERTO ESCRIVÁ[1] & YAGO RIVERA[1]
[1]Instituto de Ingeniería Energética, Universitat Politècnica de València, Spain
[2]Instituto Superior de Tecnologías y Ciencias Aplicadas, Cuba

ABSTRACT

The phenomenon of direct contact condensation of steam into pools with subcooled water is of great interest for various industrial applications, allowing rapid condensation of steam by providing high heat transfer and mass exchange capacity. For the nuclear industry it is of great interest to understand steam condensation in the presence of non-condensable (NC) gases from the point of view of passive safety of nuclear power plants. Currently there are several experimental studies related to steam and non-condensable gas discharges in pools, but work is still in progress to obtain a wider range of information. The objective of the present study is to characterize the horizontal jet behavior in the transition zone, when initially steam is being discharged and then a small fraction of air is injected. An abrupt change in behavior is observed when a jet dominated totally by momentum forces experiences the impulse of buoyancy forces induced by the non-condensable gases. This phenomenon is due to the deterioration of heat transfer caused by the presence of air. This fact limits condensation by direct contact and modifies the trajectory of the submerged gases. Direct visualization techniques using a high-speed camera and image processing methods are used to characterize this jet behavior. Different tests have been performed by varying the steam flow rate, pool water temperature and nozzle diameter. In each of the tests, the air flow rate required for the transition zone to occur was determined. The processing of the obtained images is performed by means of a multi-step subroutine in MATLAB. Experimental results showed that the water temperature and nozzle diameter play an important role in the transition zone from the pure jet to the steam-air mixture jet.
Keywords: non-condensable gas, steam, direct contact condensation, jets, two-phase flow, digital image processing.

1 INTRODUCTION

Direct contact condensation of a steam jet discharged into a water pool allows a high capacity for mass, momentum, and energy exchange from steam to water. In the industry this phenomenon is widely used, an example is the pressure suppression pool, which provides a heat sink during the actuation of the safety-relief valves, condensing the steam that may be released in the drywell during a leak or loss-of-coolant accident (LOCA), causing a rapid decrease in pressure [1]–[3].

Previous research to characterize the behavior of the steam jet has been divided into three main areas: evaluation of the shape and steam plume length, estimation of the mean heat transfer coefficient, and the development of map characterizing the condensation regime. As a summary of the contribution of these investigations, they have defined three different steam plume shapes (i.e., conical, ellipsoidal, and divergent shapes). Kerney et al. [4] development a correlation for estimating the steam plume penetration length considering a constant Stanton number (for choked injector flows), based on the effect of injector diameter, flow rate and pool temperature. Many researchers [5]–[7] built on the results obtained by Kerney and extended the ranges of experiments by considering different working fluids, flow rates and pressures. Based on pool water temperature and steam mass velocity, different regime maps were developed for low steam mass flow [8] and high steam mass flows [7]. There are several maps of steam condensation regimes in the literature. One of the most important [9] has six defined regions leading to six defined regions (i.e., Chugging (C), Transition Chugging (TC), Oscillatory Condensation (CO), Bubble Oscillatory Condensation (BCO), Stable Condensation (SC), Interfacial Oscillatory Condensation (IOC) and Non-apparent Condensation (NC)).

The measurement techniques for the study of the interface behavior are quite complex, as the interface has a highly unstable behavior. Visualization-based techniques are widely used in the study of different phenomena associated with two-phase flow. It is possible to study the trajectory of the jet, its contour, its breakup in the form of bubbles and the behavior of these bubbles. These techniques can be divided into two main groups: direct visualization techniques [10]–[24] and advanced visualization techniques [2], [25], [26]. The first is the most used and usually involves cameras incorporating charge-coupled devices (CCD) and illumination systems and the second is more complex experimentally, where laser systems are usually used as the illumination source.

As show above, previous studies have focused on the behavior of the jet discharged only of steam and very few researchers have studied steam and non-condensable gas discharges due to their complexity. In this work we will focus on characterizing the behavior that occurs when steam is discharged into subcooled water through a nozzle and a quantity of NC gas appears and how it affects the behavior of the jet.

2 EXPERIMENTAL MATERIAL AND METHODS

2.1 Experimental facility

The experimental installation allows the discharge of jets of steam and air mixture into a pool with subcooled water by using two lines. The pool is rectangular with dimensions (length × width × height) of 1,500 mm × 500 mm × 600 mm and all its walls are made of glass which allows optical measurements. The water used for both the pool and to generate steam is treated by a reverse osmosis process where all impurities and dissolved lime are removed. The water level in the pool is maintained at 500 mm and the rest of the pool height (100 mm) is considered as free surface. Fig. 1 shows in detail a schematic diagram of the two lines (steam and air), each of the equipment present in the installation, and the geometrical characteristics of the discharge pool.

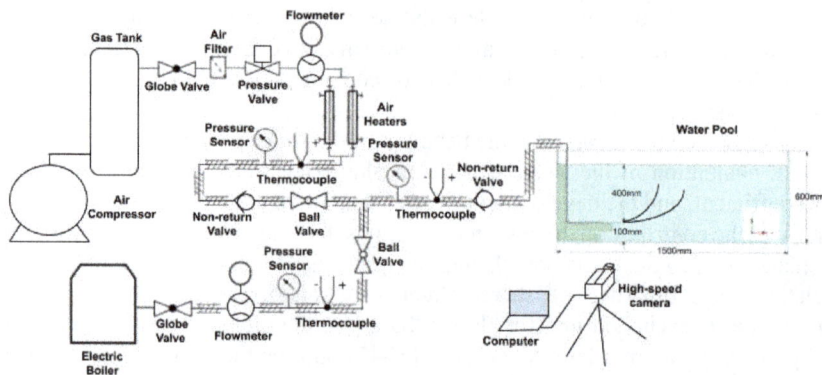

Figure 1: Schematic diagram of the experimental facility [27].

Saturated steam is obtained by an electric steam generator operating at a pressure of 6 bar. The steam volumetric flow is measured by a rotameter with an operating range of 20 to 110 l/min and a flowmeter with an operating range of 50 to 400 l/min. Downstream, a thermocouple and a pressure sensor record the parameters before the mixing zone.

The air supply line is fed by a screw air compressor with a capacity of 1,420 l/min and 10 bars of maximum operating pressure. The air is stabilized in a gas tank (boiler), which allows a better control of the discharge conditions by means of a globe valve and a precision valve. Downstream, a flowmeter allows to know the air flow rate to be discharged. Two heaters in parallel of 120 V and 500 W are installed to maintain the air temperature above the steam temperature upstream of the mixing zone to avoid condensation of the steam. The power of each heater is controlled by a power module, a closed-loop PID control system and a thermocouple used as set point. As in the steam line, a thermocouple and a pressure sensor are installed upstream of the mixing zone. To guarantee that air is perfectly mixed with the steam, the mixing line is sufficiently long enough before discharge.

To know the jet discharge parameters, a pressure sensor and a thermocouple are placed before the nozzle outlet. All sensors are connected to a National Instruments data acquisition system (Model 6259 16-bit) driven by Labview Software to monitor and control data acquisition.

The hoses are made of a flexible material that withstands temperatures up to 232°C and a pressure of 18 bar, additionally they are thermally insulated to avoid losses in the areas where the working fluids are not at room temperature.

The jet behavior is recorded by a high-speed camera (PCO.1200 hs model) with a shooting frequency of 636 frames per second (fps) at high resolution (1,280 × 1,024 pixel). To provide the necessary light for the high-speed camera, an LED light panel with dimensions 1,200 mm × 600 mm was placed at the back of the pool. Fig. 2 shows a photo taken of experimental facility.

Figure 2: Experimental facility [27].

2.2 Test matrix

For this test matrix, five interchangeable stainless-steel nozzles were used. For each steam flow rate, it was analyzed what would be the percentage of air in which it would change from a jet dominated only by the force of the momentum to a jet where buoyancy forces dominated, and a continuous curved jet would appear. The test matrix used is summarized in Table 1.

Table 1: Summary of experimental initial conditions.

Parameters	Value
Nozzle diameter	2–6 mm
Steam volumetric flux	30–20 l/min
Water temperature	25–60°C

2.3 Image analysis method

To capture the change of the jet behavior where initially there is steam until the necessary air flow is obtained where it is determined that it is a continuous jet, it is necessary to make a series of images at different percentages of air flow using the direct visualization technique by means of a high-speed camera (CCD). The method of analysis of the images has several steps. First, a background image showing only the nozzle is recorded to determine the number of pixels per millimeter as a function of the external diameter of the nozzle. Then, the discharge is recorded for each of the initial conditions defined for that nozzle. All images had a resolution of 1,024 × 1,280 pixels. According to the setting parameters defined in the camera software and the size of the RAM memory, the maximum number of images in each of the recordings of the jet discharge was 1,200 images.

The processing of these images is performed by means of a MATLAB script. The image of the jet discharge is cropped to the same size as the background image, eliminating the part that is not interesting for our study. The image in grayscale is then binarized using a specified threshold. Next, a function performs the median filtering of the image, where each output pixel contains the median value in a 3 by 3 neighborhood around the corresponding pixel. Because there are areas of the gas jet composed of some shades is applied a function for fills the holes in the input binary image. In this syntax, a hole is a set of background pixels that cannot be reached by filling in the background from the image edge (see Fig. 3(d)). Subsequently, the adjustment (morphological) operation is applied to remove small bubbles from the image, for these operations two functions are used; the first erodes the binary image and the second dilates the binary image, these functions use a disk-shaped structuring element with a radius $r = 6$ (see Fig. 3(e)). Finally, a function that returns a binary image that contains only the perimeter pixels of objects in the input image. A pixel is part of the perimeter if it is nonzero, and it is connected to at least one zero-valued pixel (see Fig. 3(f)).

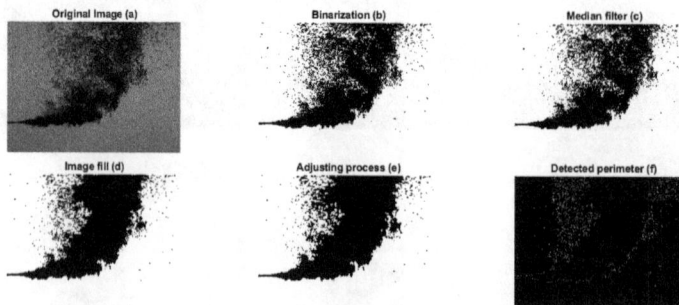

Figure 3: Steps used in the routine implemented in MATLAB to detect the jet boundary.

3 EXPERIMENTAL RESULTS AND DISCUSSION

Discharges in the form of steam jets are characterized by a practically horizontal trajectory until their extinction, behavior caused by the phenomenon of direct condensation due to the presence of water around them. While the incondensable jets present a typical curved shape, due to the balance between inertial forces and buoyancy forces. Taking these two behaviors as a premise, the present work aims to analyze how the transition between these two structures takes place. For this purpose, a progressively larger amount of non-condensable gases is introduced to a pure steam jet (Fig. 4). The main objective of this study is to determine the experimental conditions under which the transition takes place, i.e., the shape changes from a pure steam jet to a continuous jet that reaches the free sheet. Direct visualization techniques have been used to carry out this study.

Fig. 4 shows the jet behavior in each of the cases with different air fractions (V_a), taking one of the 1,200 subsequent images of each discharge series. At the first moment, when only steam was discharged (i.e., $V_s = 1$ and $V_a = 0$), the jet was barely perceptible because everything was condensing, and the momentum force predominated up to its extinction. At the beginning of the air discharge, after the air-line pressure was sufficient for the air flow to be perceptible in the jet behavior, it was possible to appreciate how some bubbles rose randomly, appearing the buoyancy phenomenon. Then, as the air volumetric fraction was increased, it can be seen how the behavior of the discharge changed from isolated bubbles to a defined steam/air-water interface.

Fig. 4 also shows how the behavior of the discharge goes from a steam condensation regime, in this case of study it was SC regime, to a regime of isolated air bubbles appearing and detaching from the jet and finally reaching the air jet regime where the buoyancy force dominates, which is because the disturbances move across the phase boundary faster than they can accumulate as noted by Liang et al. [28]. An important finding is that the transition to continuous buoyant jet does not always take place in the same way; for low flows, the transition is quite abrupt while for high flows it is more progressive.

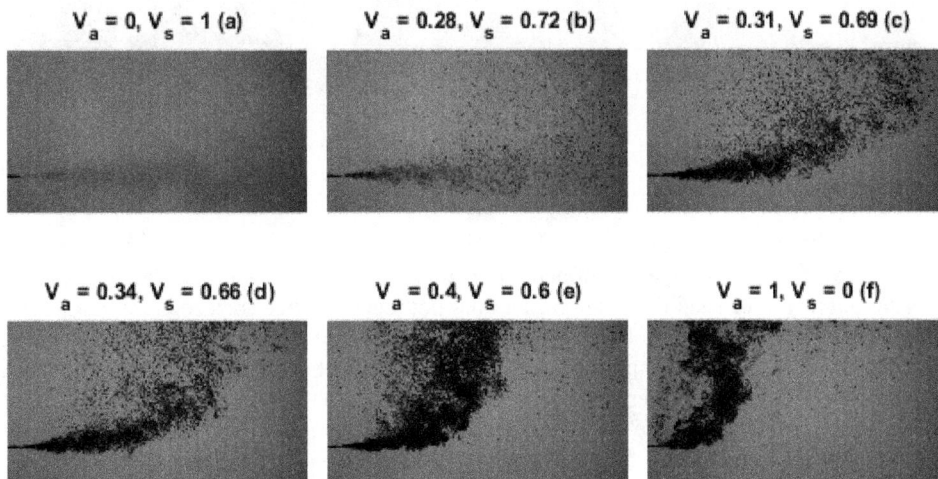

Figure 4: Images obtained for 3 mm diameter with different air volume fractions (V_a) to determine the range where the transition could occur, starting from steam volume fraction (V_s) of 1.

WIT Transactions on Engineering Sciences, Vol 132, © 2021 WIT Press
www.witpress.com, ISSN 1743-3533 (on-line)

3.1 Image of variable spatial intensity and contours

A summation method was applied to the images processed in MATLAB, in which each processed image was added to the last one until 1,200 images in each series were reached, obtaining an image with varying spatial intensity (defined by the number of images in each subsequent image series). The colors indicate how many times in all images a certain place in the field of view is occupied by gas or water. As can be seen in Fig. 5, the steam where it can be captured the most is just after the nozzle exit, the trace after that cannot be captured due to the condensation that prevents having a defined interface between the steam and the gas. Then with the increase of the air fraction the rigidity of the steam plume weakens and brings about that the volume of steam bubbles separated from the steam plume increases, which agrees with the previous results of Zhao et al. [29].

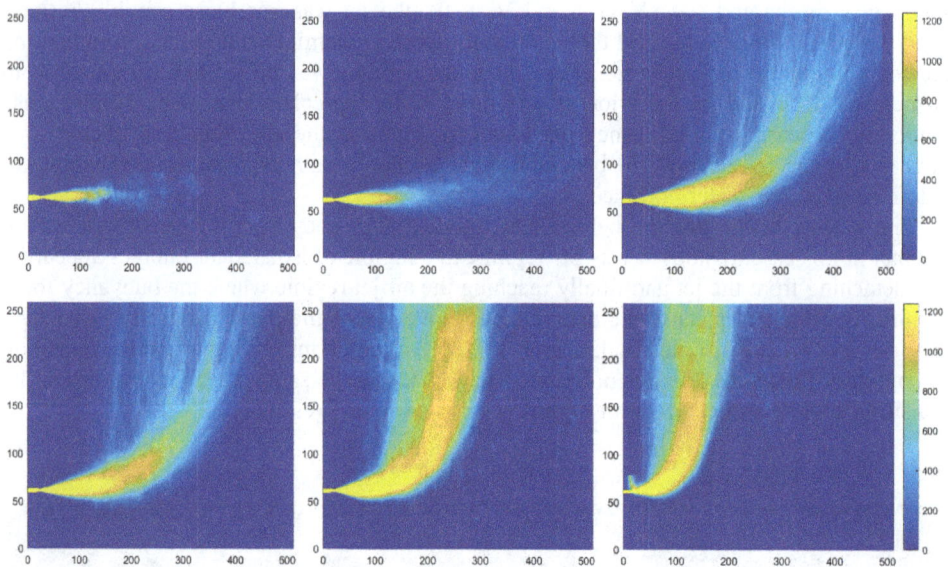

Figure 5:   Images of variable spatial intensity obtained for 3 mm diameter with different air volume fractions.

Fig. 5(c) and (d) show a more defined plume where a more noticeable curvature is already present, but as can be seen the steam/air jet does not reach the free surface in all images due to the pinch-off phenomenon. In addition, the steam/air plume lengthens, and the gas-water interface expands, which is in agreement with that proposed by Xu [30]. The transition zone was defined as the spatial intensity image in which there was continuity of the jet from the nozzle exit to the free surface and the pinch-off did not appear in the whole series of the processed images (Fig. 5(e)).

To show more clearly where both gas and water are located, the variable spatial intensity image in Fig. 5 is delimited by filled contour isolines (Fig. 6). These isolines delimit the different levels, in this case 6 levels were used, representing in yellow the area where the gas is in most of the images. In these images it can be seen how the large amount of steam that was at the outlet of the nozzle and formed the steam cavity will be dragged by the air current.

As can be seen in the figures when we had a considerable amount of air, the gas-water interface expanded sharply, and the mixture layer thickened. The air layer at the interface prevented the direct contact condensation by increasing the condensation resistance, which corroborates the work of Xiaoping et al. [31].

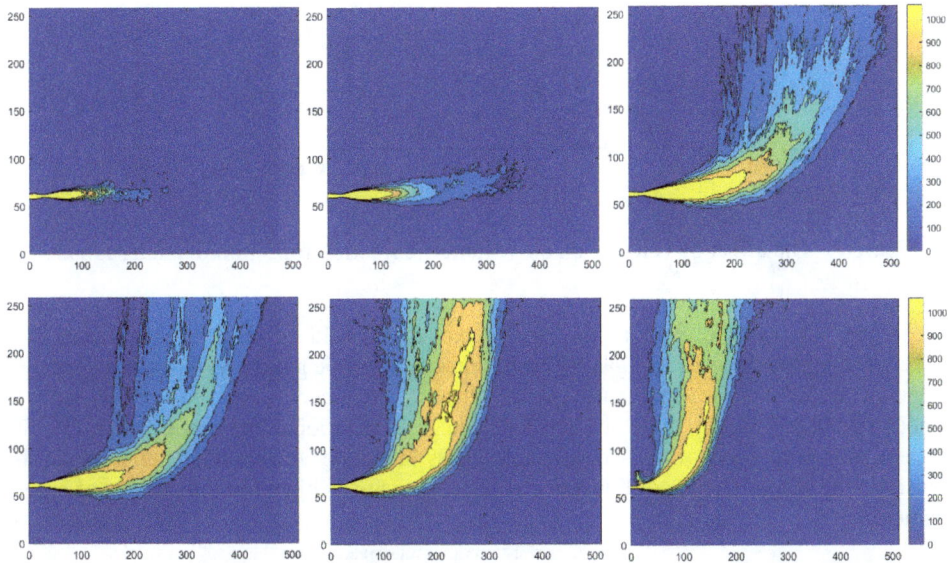

Figure 6:   Images of filled contour isolines obtained for 3 mm diameter with different air volume fractions.

3.2 Dependence of the velocity on air volumetric fraction

For the same nozzle diameter, the influence of velocity on the air volumetric fraction required to reach the transition was studied. As can be seen in Fig. 7, velocity was not found to influence the air volume fraction, i.e., there is not most influence of the mass flow rate. This is believed to be because there are several phenomena that counteract each other and do not have a predominant one. For example, as the velocity increases, there is a greater initial expansion (since there is a higher inlet pressure, then a more abrupt expansion has to occurs to equalize its pressure with that of the surrounding medium), consequently a more superheated steam, which makes condensation more difficult and facilitates that a smaller amount of air is able to pull the steam, causing a continuous upward plume to form; but at the same time, there is a greater expansion angle and a higher degree of turbulence which causes greater heat transfer, favoring the steam condensation.

3.3 Dependence of the cross-sectional area on air volumetric fraction

An important point that was investigated was how the cross-sectional area for each of the nozzles affected the air volume fraction. As can be seen in Fig. 8, as the diameter increases and with it the cross-sectional area, a lower percentage of air is needed to reach the transition, as shown in the previous section there is not a clear influence of the mass flow rate of steam

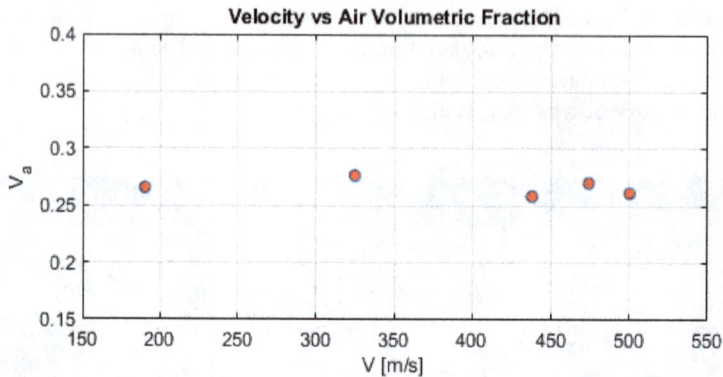

Figure 7: The effect of velocity on air volumetric fraction for 3 mm diameter nozzle.

discharged. As possible explanation for this tendency could be because when there is a presence of mixed flows of two gases (steam/non-condensable gases), as the steam condense in the gas/liquid interface then consequently there is a tendency of the non-condensable to concentrate in this region, these non-condensable gases deteriorate the heat transfer. Despite of the major proportion of non-condensable gases in smaller nozzle diameter to reach the transition the absolute value of this non-condensable gases is higher in the nozzles of bigger diameter.

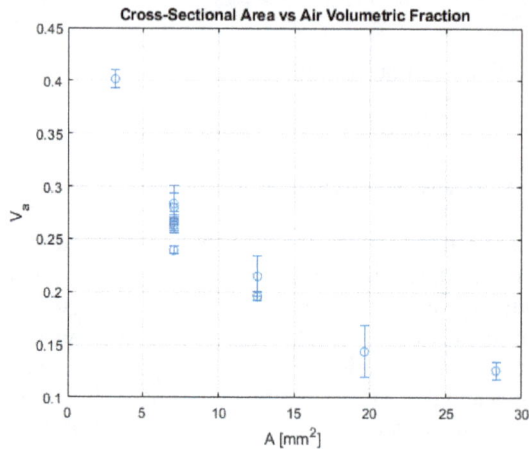

Figure 8: The effect of cross-sectional area on air volumetric fraction for all nozzles diameters.

4 CONCLUSIONS

The behavior of steam discharge in stagnant water when air is injected into it is experimentally investigated and it is determined when the transition from steam jet to rising plume jet occurs. The main conclusions obtained are:

1. The variable spatial intensity image obtained for each of the cases allowed determination of the range of air volumetric fraction necessary for the transition to occur.

2. No dependence on increasing velocity was obtained that would bring an abrupt change in air volume fraction for the same diameter.
3. A dependence between nozzle cross-sectional area and air volume fraction was obtained, i.e., as the cross-sectional area increased, a smaller air volume fraction was necessary to obtain the transition, independent of the mass flow rate of steam discharged.

5 FUTURE WORKS

Due to the fact that this work is a first approach to the subject of the transition from pure steam discharge to rising plume discharge, it will be considered for future work to perform a greater number of tests to expand the data base, where a greater number of discharge velocities, various pool water temperatures, nozzle diameters, etc. will be taken into account. It is also considered to study the contribution of other variables in this phenomenon and to obtain a correlation of some parameters such as air volume fraction, expansion angle, nozzle diameter and pool water temperature.

ACKNOWLEDGEMENTS

The authors are indebted to the support of the Spain plan of I + D support to the EXMOTRANSIN project ENE2016-79489-C2-1-P and the Santiago Grisolía Program for its training research personnel.

REFERENCES

[1] Norman, T.L. & Revankar, S.T., Jet-plume condensation of steam-air mixtures in subcooled water, part 1: Experiments. *Nucl. Eng. Des.*, **240**(3), pp. 524–532, 2010.
[2] Song, D., Experimental and numerical investigation of thermal stratification by direct contact condensation of steam in pressure suppression pool. The University of Tokyo, 2014.
[3] Rassame, S. et al., Experimental investigation of void distribution in suppression pool during the initial blowdown period of a loss of coolant accident using air-water two-phase mixture. *Ann. Nucl. Energy*, **73**, pp. 53–67, 2014.
[4] Kerney, P.J., Faeth, G.M. & Olson, D.R., Penetration characteristics of a submerged steam jet. *AICHE J.*, **18**(3), pp. 548–553, 1972.
[5] Weimer, J.C., Faeth, G.M. & Olson, D.R., Penetration of vapor jets submerged in subcooled liquids. *AICHE J.*, **19**(3), pp. 552–558, 1973.
[6] Kim, Y.S., Park, J.W. & Song, C.H., Investigation of the stem-water direct contact condensation heat transfer coefficients using interfacial transport models. *Int. Commun. Heat Mass Transf.*, **31**(3), pp. 397–408, 2004.
[7] Chun, M.-H., Kim, Y.-S. & Park, J.-W., An investigation of direct condensation of steam jet in subcooled water. *Int. Commun. Heat Mass Transf.*, **23**(7), pp. 947–958, 1996.
[8] Chan, C.K. & Lee, C.K.B., A regime map for direct contact condensation. *Int. J. Multiph. Flow*, **8**(1), pp. 11–20, 1982.
[9] Cho, M.K., Song, S., Park, C.H., Yang, C.K. & Chung, S.K., Experimental study on dynamic pressure pulse in direct contact condensation of steam jets discharging into subcooled water. NTHAS98 1. *Korea-Japan Symp. Nucl. Therm. Hydraul. Saf.*, 1998.
[10] Shi, H.H., Wang, B.Y. & Dai, Z.Q., Research on the mechanics of underwater supersonic gas jets. *Sci. China Physics. Mech. Astron.*, **53**(3), pp. 527–535, 2010.
[11] Harby, K., Chiva, S. & Muñoz-Cobo, J.L., An experimental investigation on the characteristics of submerged horizontal gas jets in liquid ambient. *Exp. Therm. Fluid Sci.*, **53**, pp. 26–39, 2014.

[12] Li, W., Wang, J., Sun, Z., Zhou, Y., Liu, J. & Meng, Z., Experimental investigation on thermal stratification induced by steam direct contact condensation with non-condensable gas. *Appl. Therm. Eng.*, **154**, pp. 628–636, 2019.

[13] Li, W., Wang, J., Zhou, Y., Sun, Z. & Meng, Z., Investigation on steam contact condensation injected vertically at low mass flux: Part I pure steam experiment. *Int. J. Heat Mass Transf.*, **131**, pp. 301–312, 2019.

[14] Xu, Q., Liu, W., Li, W., Yao, T., Chu, X. & Guo, L., Experimental investigation on interfacial behavior and its associated pressure oscillation in steam jet condensation in subcooled water flow. *Int. J. Heat Mass Transf.*, **145**, 2019.

[15] Jo, B., Erkan, N. & Okamoto, K., Richardson number criteria for direct-contact-condensation-induced thermal stratification using visualization. *Prog. Nucl. Energy*, **118**, 2020.

[16] Xu, Q., Li, W., Chang, Y., Yu, H. & Guo, L., A quantification of penetration length of steam jet condensation in turbulent water flow in a vertical pipe. *Int. J. Heat Mass Transf.*, **146**, 2020.

[17] Qu, X.H., Tian, M.C., Zhang, G.M. & Leng, X.L., Experimental and numerical investigations on the air-steam mixture bubble condensation characteristics in stagnant cool water. *Nucl. Eng. Des.*, **285**, pp. 188–196, 2015.

[18] Song, D., Erkan, N., Jo, B. & Okamoto, K., Relationship between thermal stratification and flow patterns in steam-quenching suppression pool. *Int. J. Heat Fluid Flow*, **56**, pp. 209–217, 2015.

[19] Xu, Q. & Guo, L., Direct contact condensation of steam jet in crossflow of water in a vertical pipe. Experimental investigation on condensation regime diagram and jet penetration length. *Int. J. Heat Mass Transf.*, **94**, pp. 528–538, 2016.

[20] Qu, X.-H. & Tian, M.-C., Acoustic and visual study on condensation of steam-air mixture jet plume in subcooled water. *Chem. Eng. Sci.*, **144**, pp. 216–223, 2016.

[21] Harby, K., Chiva, S. & Muñoz-Cobo, J.L., Modelling and experimental investigation of horizontal buoyant gas jets injected into stagnant uniform ambient liquid. *Int. J. Multiph. Flow*, **93**, pp. 33–47, 2017.

[22] Li, S.Q., Lu, T., Wang, L. & Chen, H.S., Experiment study on steam-water direct contact condensation in water flow in a Tee junction. *Appl. Therm. Eng.*, 2017.

[23] Li, W., Meng, Z., Wang, J. & Sun, Z., Effect of non-condensable gas on pressure oscillation of submerged steam jet condensation in condensation oscillation regime. *Int. J. Heat Mass Transf.*, **124**, pp. 141–149, 2018.

[24] Li, W., Meng, Z., Sun, Z., Sun, L. & Wang, C., Investigations on the penetration length of steam-air mixture jets injected horizontally and vertically in quiescent water. *Int. J. Heat Mass Transf.*, **122**, pp. 89–98, 2018.

[25] Dahikar, S.K., Sathe, M.J. & Joshi, J.B., Investigation of flow and temperature patterns in direct contact condensation using PIV, PLIF and CFD. *Chem. Eng. Sci.*, **65**(16), pp. 4606–4620, 2010.

[26] Choo, Y.J. & Song, C.-H.H., PIV measurements of turbulent jet and pool mixing produced by a steam jet discharge in a subcooled water pool. *Nuclear Engineering and Design*, **240**(9), pp. 2215–2224, 2010.

[27] Cordova, Y., Rivera, Y., Blanco, D., Berna, C., Muñoz-Cobo, J.L. & Escrivá, A., Experimental investigation of submerged horizontal air–steam mixture jets into stagnant water. *Adv. Fluid Mech. XIII*, **1**, pp. 89–101, 2020.

[28] Liang, K.S. & Griffith, P., Experimental and analytical study of direct contact condensation of steam in water. *Nucl. Eng. Des.*, **147**(3), pp. 425–435, 1994.

[29] Zhao, Q., Cong, Y., Wang, Y., Chen, W., Chong, D. & Yan, J., Effect of non-condensation gas on pressure oscillation of submerged steam jet condensation. *Nucl. Eng. Des.*, **305**, pp. 110–120, 2016.

[30] Xu, Y., Direct contact condensation with and without non condensing gas in a water pool. Purdue University, West Lafayette, 2004.

[31] Xiaoping, Y., Daotong, C., Jiping, L., Xiao, Z. & Junjie, Y., Experimental study on the direct contact condensation of the steam-air mixture in subcooled water flow in a rectangular channel. *Int. J. Heat Mass Transf.*, **88**, pp. 424–432, 2015.

COMPUTATIONAL INVESTIGATION OF THE EFFECT OF THE ASPECT RATIO ON SECONDARY CURRENTS IN OPEN CHANNELS

MONSIF SHINNEEB[1], GHASSAN NASIF[1,2] & RAM BALACHANDAR[2]
[1]Department of Mechanical Engineering, Higher Colleges of Technology, UAE
[2]Department of Mechanical Engineering, University of Windsor, Canada

ABSTRACT

This study investigates numerically the effect of the aspect ratio (AR) on the velocity field characteristics of the turbulent flow of a straight open channel flow. The AR is defined as the ratio of the width of the channel in a plane normal to the flow direction, to the flow depth. In this study, two aspect ratio cases are investigated; a narrow case of AR = 1 and a wide case of AR = 12. The transient three-dimensional Navier–Stokes equations were numerically solved using a finite-volume approach with detached-eddy simulation (DES) turbulence model. The free surface was simulated using flat-wave model linked with the volume of fluid method. The objective of this study is to enhance our understanding of the effect of the AR on the formation of secondary currents in a channel flow. The results revealed the formation of a pair of counter-rotating recirculation zones near the bottom corners of the channel, whose axes are aligned with the main flow direction. The AR appears to significantly influence the size and strength of the recirculation zone that resides near the sidewalls. The distribution of the turbulent kinetic energy of the flow appears quantitatively similar for both the AR cases; however, the magnitude appears to increase with decreasing the AR.
Keywords: secondary currents, turbulence, DES model.

1 INTRODUCTION

Turbulent flow is a very complex phenomenon it is one of the most challenging topics in fluid mechanics. Moreover, the presence of solid walls bounding the flow causes the formation of boundary layer, which makes the flow three-dimensional (3D), and thus result in a more intricate flow field due to the wall effect. One of the interesting phenomenon in turbulent flow is the formation of secondary currents. The strength of the secondary currents in the flow field depends on different parameters such as the aspect ratio (AR), the flow depth, wall roughness, the distance from the solid boundary, Reynolds number, etc. This phenomenon is usually observed in channel and duct flows. Therefore, understanding the mechanism of these currents is crucial since they affect the characteristics of turbulence and mean flow.

Open-channel flow is a suitable case to understand the characteristics of the secondary currents. Generally, there are two standard types of secondary currents classified in open-channels [1]. The secondary currents that occur in a straight open channel flow due to turbulence phenomenon is known as Prandtl's second kind, which is the focus of this study. It is reported that the second type of secondary currents occurs due to turbulence anisotropy between velocity components [2]. Nezu and Rodi [3] carried out an experimental study on secondary currents in smooth open-channel flows using a laser-Doppler anemometer (LDA). It was concluded that the secondary current patterns are different from those of duct flows due to the presence of the free-surface. Nezu et al. [4] investigated the formation of secondary currents in a smooth rectangular open channel flow by varying the aspect ratio. It was reported that the secondary currents develop along the corner bisector (45°-line) regardless of the aspect ratio (AR), and it produces a pair of recirculation zones. One recirculation zone resides above the corner bisector-line which is called the "side-vortex," and the other one resides under the bisector-line which is called the "bottom-vortex." The side- and bottom-vortices are symmetrical with each other with respect to the bisector for aspect ratio AR = 2.0 [4]. The size of the side-vortex is completely restricted by the sidewall, and therefore it is not often affected by the AR of the open-channel. On the other hand, the bottom-vortex is greatly influenced by the AR. As AR increases, the size of the bottom-vortex expands and then tends to reach a constant size at larger aspect ratios. Nezu and Nakagawa [5] provided an explanation of the dependence of the secondary currents on the channel AR. In narrow channels (AR < 5), the sidewalls and bed cause an increased anisotropy between fluctuating velocity components, which result in strong secondary currents over the entire channel cross-section. For AR > 10, it is argued that the sidewall effects die out and the flow becomes two-dimensional in the central/core region of the channel. Tominaga et al. [6] carried out an experimental study using a hot-film anemometer (HWA) to investigate the secondary currents in rectangular and trapezoidal cross-section open-channel flows. It was concluded that the secondary currents are generated as a result of the anisotropy of turbulence caused by the boundary conditions, and that the secondary currents affect the mean streamwise flow, producing three-dimensional flow.

Even though the secondary currents in turbulent flow have been extensively investigated by numerous researchers, there has still been considerable controversy between researchers about the mechanism of the flow and the influence of the geometry parameters such as the aspect ratio on the formation of the secondary currents. This may be attributed to the three-dimensional nature of the turbulent flow and the inherent limitations and high cost of conventional experimental methods, which limits their effectiveness in such studies compared to numerical approaches. The availability of enhanced computational tools makes the investigation of the secondary currents in the flow field more efficient. Thus, the objective of this study is to enhance our understanding of the effect of the AR on the formation of secondary currents in a smooth open-channel flow using a numerical approach. This is achieved by contrasting the mean velocity field and turbulence characteristics at low and high AR. Two aspect ratio cases are investigated in this study; a narrow case of AR = 1 and a wide case of AR = 12. A three-dimensional time-dependent detached eddy simulation (DES) turbulent model is used in the study. The free surface was simulated using flat-wave model linked with the volume of fluid method. The aim is to provide an enhanced analysis hitherto that is not possible by experimental approaches. The importance of this study emerges from our need to gain a better understanding of secondary currents, with an ultimate goal to permit better management of the dynamic flow features in a variety of engineering and/or environmental applications.

2 MODEL SETUP AND BOUNDARY CONDITIONS

A schematic diagram of the open channel, the computational domain details, and the relevant boundary conditions that have been employed in this investigation are shown in Fig. 1. The aspect ratio (AR), which is the main parameter in this study, is determined by the ratio of the horizontal width of the rectangular channel cross-section to the water layer depth. As shown in Fig. 1(a), the computational domain consists of two regions; the water (represented by red cut lines) and air (represented by blue dots). The height (H) and width (B) of both water and air regions in the channel are the same for all cases used in the study. The channel streamwise length is kept constant for all cases, i.e. $L = 500$ mm. The water layer depth $H = 30$ mm, and the two channel widths $B = 30$ mm and 360 mm, are used to produce the channel aspect ratios AR = 1 and 12, respectively. The inlet velocity at the channel inlet boundary is uniform for all cases in this study with a magnitude equal to 0.75 m/s. All wall boundaries in contact with

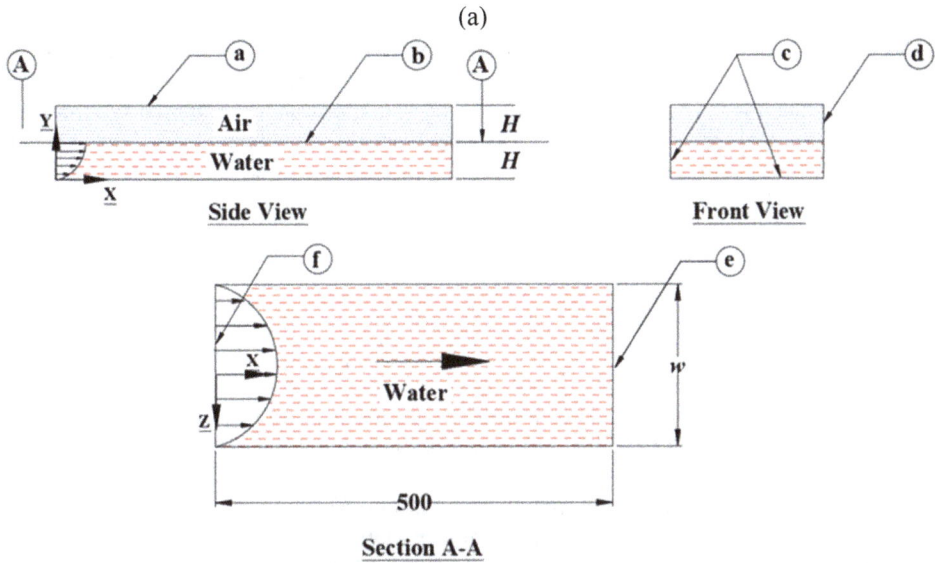

(a) Pressure outlet; (b) Water-air interface; (c) No-slip wall (water); (d) Slip wall (air); (e) Outlet boundary; and (f) Velocity inlet.

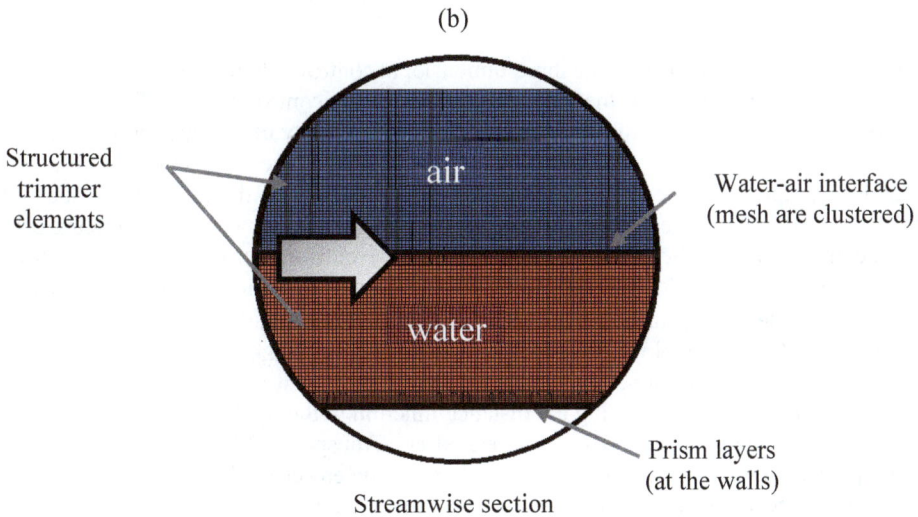

Figure 1:   (a) A computational model with appropriate boundary conditions; and (b) Section through the computational domain in the streamwise direction [8].

the water region are considered as a no-slip smooth boundary, while all sidewalls that are in contact with the air region are considered as a slip boundary. The top surface of the air (labelled with letter "a" in Fig. 1) is treated as a pressure outlet boundary. The hydrostatic pressure of the flat wave model [7] is used as a boundary for the pressure, and the outflow boundary is used for the velocity in the water region at the outlet (labelled with letter "e" in Fig. 1). To reduce the effect of the numerical reflection of the waves from the boundaries, a damping factor is used for this purpose. The fully-developed turbulent velocity profile, which is used as an inlet boundary condition (labelled with letter "f" in Fig. 1), is obtained by periodically mapping the velocity components and turbulent fluctuations distribution from the y-z plane (see Fig. 1) located mid-way ($x/L = 0.5$) in the streamwise direction x, until the mean velocity profiles become almost identical across the channel flow at different streamwise locations [8]. The reason for extracting the data at this streamwise location rather than at the outlet boundary is to avoid the influences of the exit boundary condition since the velocity is observed to increase as it approaches the channel exit due to a slight decline in the water level.

3 COMPUTATIONAL METHODOLOGY

The open channel flow is simulated using Siemens PLM's STAR-CCM+ [9] with a structured trimmer mesh. The trimmer meshing model utilizes a template mesh that is constructed from hexahedral cells from which it cuts or trims the core mesh using the structured hexahedral surface mesh as the starting input [9]. First-order implicit marching in time and second-order differencing in space are used to discretize the governing equations. The time-dependent governing equations comprise a continuity equation for conservation of mass, three conservation of momentum equations, and conservation of energy equation. Each of these equations can be expressed in a general form by the transport of a specific scalar quantity ϕ per unit mass, represented in a continuous integral form as [10]:

$$\frac{\partial}{\partial t}\int_{CV}\rho\phi dV + \oint_A \mathbf{n}\cdot(\rho\phi\mathbf{u})dA = \oint_A \mathbf{n}\cdot(\Gamma_\phi\nabla\phi)dA + \int_{CV}S_\phi\,dV \tag{1}$$

where CV in eqn (1) represents the three-dimensional control volume over which the volume integration is carried out, A is the bounding surface of the control volume. The terms in eqn (1) from left to right are the rate of change of the total quantity of the fluid property ϕ in the control volume, the rate of change of the property ϕ due to the convection flux across the bounding surface of the control volume, the rate of change of the property ϕ due to the diffusive flux across the bounding surface of the control volume, and the volumetric source in the control volume. The unit vector $\mathbf{n}$ in eqn (1) is the outward normal vector to the surface, $\mathbf{u}$ is the instantaneous velocity vector, ρ is the density, and Γ_ϕ is the diffusion coefficient.

Improved delayed detached-eddy simulation (IDDES) model [11]–[13] is a modified turbulent model that employs Reynolds-Averaged Navier–Stokes (RANS) equation at the near-wall regions and Large-eddy simulation (LES) at the rest of the flow. The model was originally formulated by replacing the distance function in the Spalart–Allmaras model with a modified distance function. The k-ε Shear Stress Transport (SST) turbulence model is a two-equation eddy-viscosity model [14], [15] and has been selected as the RANS part of the DES turbulence model in this study. The k-ε SST model solves additional transport equations for turbulent kinetic energy k and specific dissipation rate ε, from which the turbulent kinematic viscosity ($v_t = k/\varepsilon$) can be derived. The transport equations of k and ε are described in [15].

Since the flow field under investigation involves two immiscible fluids, a numerical model to handle two-phase flow is required. The volume of fluid (VOF) model [16] is a simplified multiphase approach that is well suited to simulate flows of several immiscible fluids and is capable of resolving the interface between the mixture phases. In a VOF simulation, the basic model assumption is that all phases share the same velocity and pressure and no additional modelling of inter-phase interaction is required. The normalized variable diagram provides the methodology used in constructing high-resolution schemes [17]. The

Compressive Interface Capturing Scheme for Arbitrary Meshes (CICSAM) [18] and the High-Resolution Interface Capturing Scheme (HRIC) [19] are the most commonly used high-resolution schemes for interface capturing with the VOF model. The HRIC scheme is used to capture the interface in the present work. In high-resolution schemes, an additional condition is required, the convective boundedness criterion must be satisfied along with the local Courant–Friedrichs–Lewy (CFL) condition. The CFL condition is a necessary condition for numerical stability. If an explicit time marching solver is used, then a Courant number less than one is typically required. Implicit solvers, like the one used in this study, are less sensitive to numerical instability and larger values of Courant number may be tolerated [9].

4 MESH GENERATION

In the present study, structured-trimmer elements are used to mesh the computational domain. Preliminary open-channel flow simulations utilizing the flow field that resembles experimental studies, which are carried out in a recirculating open channel flume at the Hydraulic Engineering Research Laboratory at the University of Windsor [20], were performed first for cell sensitivity inspection and model validation purposes. Detailed validations were carried out by comparing the streamwise velocity and root-mean-square of turbulence level profiles from experimental and computational results. In each case, flow field parameters were checked and compared with experimental results. Furthermore, successive grids were compared to determine whether or not there was a change in the mean characteristics. The results for the mean flow quantities and cell independence study were very similar to those reported earlier in [8], [21], [22] and are not repeated here for brevity. In this study, the total number of cells used for the computation is 1.3 and 8.7 million elements for AR = 1 and 12, respectively. In the present study, ten layers of fine prism cells, packed within 1.5 mm using a stretching factor of 1.5, are employed to resolve the wall effect. The non-dimensional wall-normal distance (y^+) value is less than two everywhere in the entire computational domain, which lies in the viscous sub-layer. The cells are also clustered along the water-air interface to reduce the numerical diffusion and preserve the sharpness of the water-air interface. To select the proper time step (Δt), different time steps were tried to satisfy the local Courant–Friedrichs–Lewy (CFL) condition [9]. The final time step was set as 2×10^{-4} s, yielding a Courant number less than 0.5 in the entire computational domain. Five internal iterations were employed at each time step. The physical time that is used for averaging the transient quantities is 15 s for all cases of this investigation. This time is initiated after the instantaneous flow field reaches its steady condition. The numerical results are considered to have converged when the scaled continuity and momentum residuals fall below 10^{-6}.

5 RESULTS

In this section, vector plots and/or color contours of the mean velocity field of an open channel flow are presented and discussed. Two cases are discussed in this paper to highlight the effect of aspect ratio (AR) on the formation of secondary currents; one represents a small aspect ratio (AR = 1) and the other one represents a much larger aspect ratio (AR = 12) which is used as a reference case. In the following results, x-, y-, and z-axes represent streamwise, vertical, and horizontal locations respectively; while U, V, and W represent the corresponding mean velocity components.

5.1 Boundary layer characteristics

The purpose of this section is to provide an extra validation of the current computational results of the open channel flow, and also to document the characteristics of the boundary layer flow. To confirm that the current results conform to the published results, a comparison between the streamwise velocity U profile in the fully-developed region for the AR = 12 case and well-documented channel flow results is made. The bulk velocity U_b of the channel flow in this study is 0.75 m/s. The current results displayed that the velocity profile of the channel flow varies throughout the water layer depth H, which suggests that the boundary layer can be assumed to extend to the free surface [23], [24]. However, the boundary layer thickness δ (perpendicular to the bed) in this study is estimated based on $0.995U_b$, and found to occupy ~29.3% of the water layer depth H. Consequently, the displacement thickness δ^* and momentum thickness θ were estimated to be 0.58 and 0.41 mm, respectively. The resulting shape factor is 1.41 and the Reynolds number based on the momentum thickness Re_θ is ~308.

Fig. 2 shows distribution of the streamwise component of the mean velocity in the channel flow in terms of the dimensionless wall velocity $U^+ (= u/u_\tau)$ and vertical axis $y^+ (= yu_\tau/v)$. The friction velocity u_τ, defined as $\sqrt{\tau/\rho}$, was determined by the Clauser chart method [25], which is based on the assumption that the velocity profile follows a universal logarithmic form in the overlap region of the boundary layer. In this study, the friction velocity u_τ was estimated to be 39 mm/s. The corresponding skin friction coefficient C_f, defined by $2(u_\tau/U_b)^2$, was found to be 5.41×10^{-3}. The viscous length scale l_v, defined by $(v/u_\tau$, where v is the kinematic viscosity of the water) was estimated to be 0.026 mm. The present velocity profile was also compared with the experimental results of Balachandar and Bhuiyan [26] for smooth channel flow at a similar Reynolds number. The present results are in good agreement with the previous channel flow data.

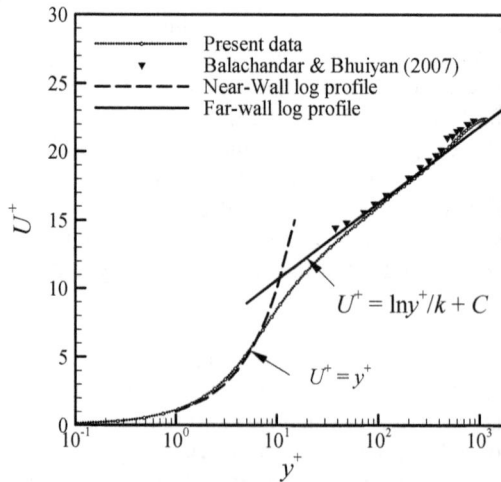

Figure 2: Mean streamwise velocity distribution of the smooth open-channel flow using the inner coordinate. The current velocity data for AR = 12 is compared with experimental results.

5.2 Mean velocity field

Fig. 3 shows two vector plots of the mean velocity field in *y-z* plane for aspect ratios AR = 1 and 12, respectively, extracted from the fully-developed flow region of an open channel flow. In these plots, the colour contour, which represents the mean streamwise velocity component, is also shown in the plots to provide a better description of 3D velocity field. In this figure, locations are normalized by the water layer depth *H* and velocities are normalized by the bulk velocity U_b. Note that only some vectors are shown to avoid cluttering on the figures. The vector plot shown in Fig. 3(a) clearly illustrates the formation a pair of strong counter-rotating recirculation zones near the bottom corners of the channel for the AR = 1 case, whose axes are aligned with the streamwise direction *x*; one of the pair resides near the bottom bed while other one is relatively bigger and resides near the sidewalls. Furthermore, there is a strong downward flow in the mid-vertical plane driven by the recirculation zones that reside near the sidewalls which meets with a relatively weaker upward flow driven by the lower recirculation zones. Fig. 3(b) shows that the behavior of the AR = 12 case is very different. It shows the formation of a pair of counter-rotating recirculation zones in the bottom corners of the channel. However, the strengths of the recirculation zones near the bed are quite energetic and they appear to span almost the whole flow depth, while the zones that reside near the sidewalls appear weak and confined in a narrow region near the sidewalls (see the region highlighted by a red square). In the core region of this wide channel (AR = 12), Fig. 3(b) shows that the magnitude of the mean vertical *V* and horizontal *W* velocity components

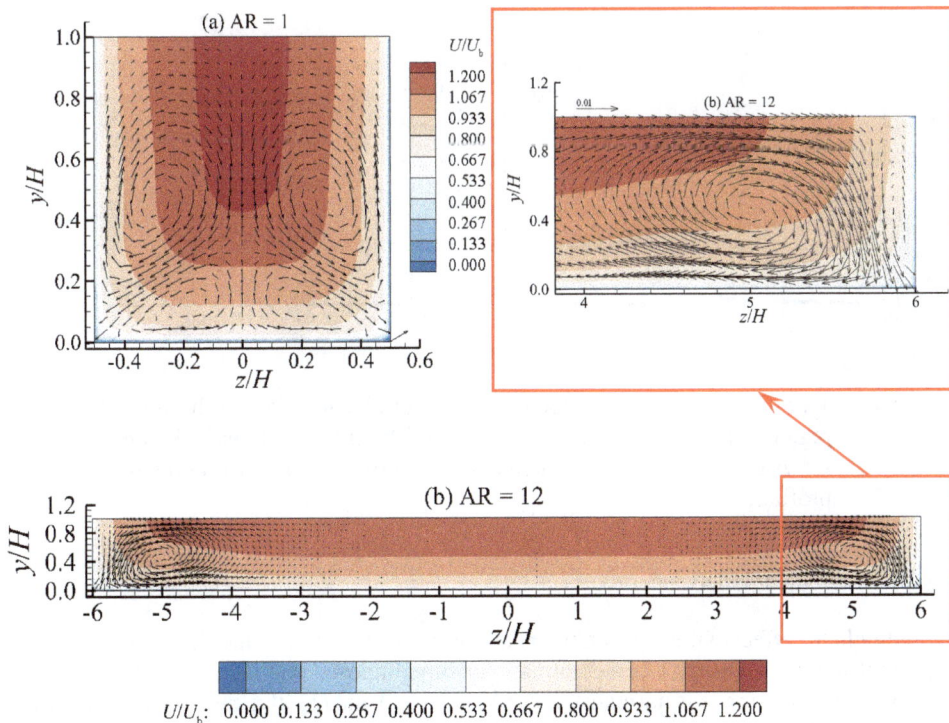

Figure 3:   Mean velocity field for an open channel flow in the fully-developed flow region for (a) AR = 1; and (b) AR = 12. The colour contour represents the normalized mean streamwise velocity U/U_b component.

are almost zero. On the other hand, the colour contour generally illustrates that the mean streamwise velocity is zero at the solid walls and increase gradually to reach its maximum magnitude in the core region of the channel flow for AR = 1 and 12 as shown in Figs 3(a) and 3(b). However, the darker brown colour in the core region in Fig. 3(a) compared to Fig. 3(b) indicates that the magnitude of U for AR = 1 is larger than the AR = 12 case. This may be attributed to formation of the boundary layers on the solid bottom and side walls, which displaces the flow vertically and horizontally, respectively, and thus enhance U.

To quantify the effect of the AR on the velocity field, the mean streamwise U and vertical V velocity profiles along the vertical mid-plane ($z/H = 0$) of the flow depth are presented in Fig. 4. In this figure, U and V are normalized by the bulk velocity U_b and the vertical locations y by the flow depth H. As shown in Fig. 4(a), U/U_b for the AR = 1 case appears larger than the AR = 12 case. The maximum magnitude of U/U_b for the AR = 1 and 12 is ~1.24 and ~1.16, respectively. Moreover, Fig. 4(b) illustrates a relatively strong downward flow (negative V/U_b) from the free surface towards the bed, and an upward flow (positive V/U_b) originated near the bed for AR = 1. The maximum magnitude of the downward and upward flows is $|V/U_b| \approx 0.008$ and 0.005, respectively. On the other hand, V/U_b is zero for the AR = 12 throughout the flow depth. For the mean horizontal velocity component (W/U_b), it is obviously zero at the vertical mid-plane for both cases because of the symmetry.

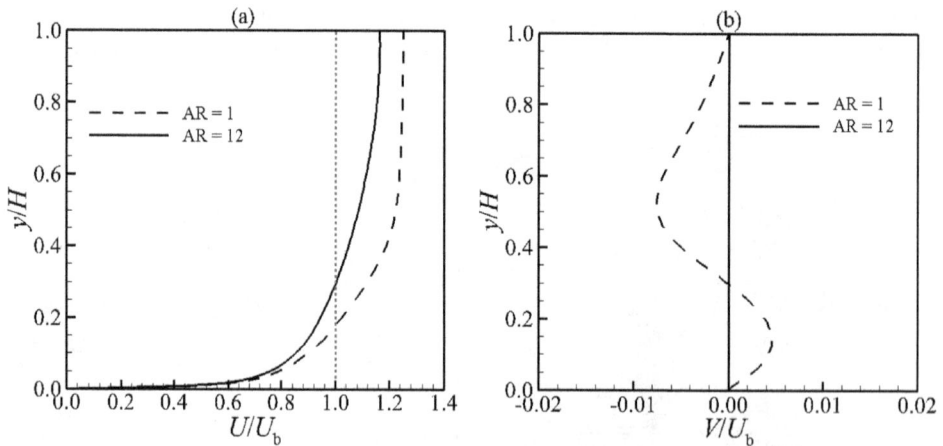

Figure 4:   Variation of the mean velocity profiles of channel flow in the fully-developed region with the flow depth y/H extracted from the mid-vertical plane ($z/H = 0$) for AR = 1 and 12. (a) Streamwise U profiles; and (b) Vertical V velocity profiles.

5.3 Mean turbulent kinetic energy

To highlight the effect of the aspect ratio on the energy distribution in the flow, Fig. 5 displays the turbulent kinetic energy k as a color contour superimposed on the mean velocity field in y-z plane. In this figure, k is normalized by U_b^2. Fig. 5(a) and (b) illustrate that the maximum turbulent kinetic energy k occurs in the layer adjacent to the bottom and side walls (brown color), while the minimum k occurs in the core region of the flow far from the solid walls (blue color). To have a better perception of the effect of the AR, k profiles are shown in Fig. 6 at horizontal locations $z/B = 0$, 0.25, and 0.42. Here B represents the width of the channel.

Note that $z/B = 0$ represents the vertical mid-plane and $z/B = \pm 0.5$ represents the location of the sidewalls. Consistent with Figs 5 and 6(a)–(c) show that the peak value of k is located near the bottom solid wall at $y/H \approx 0.016$ with a magnitude of $k/U_b^2 \approx 0.014 \pm 0.001$ for all z/B locations. Far from the bed, the magnitude of k is generally smaller over most of the flow depth for the AR = 1 case compared to the AR = 12 case at $z/B = 0$. However, the magnitude of k appears to slightly increase for the AR = 12 near the free surface as it approaches the sidewalls, but the rate of increase of k for the AR = 1 is much larger until it becomes almost uniform at $z/B = 0.42$ with a magnitude of $k/U_b^2 \approx 0.01$.

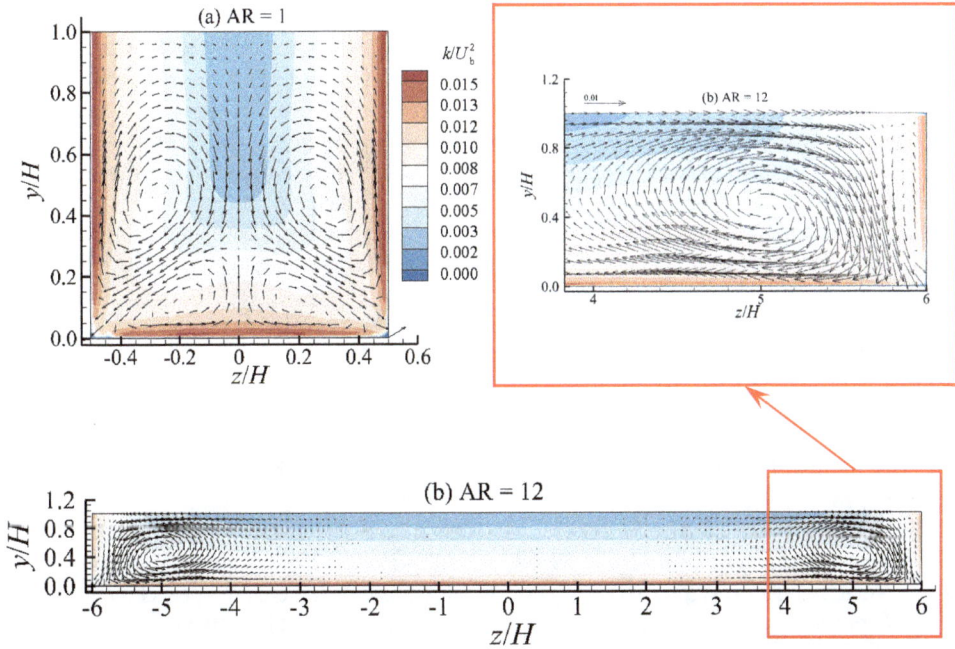

Figure 5:   Mean velocity field for an open channel flow in the fully-developed flow region for (a) AR = 1; and (b) AR = 12. The colour contour represents the normalized turbulent kinetic energy k/U_b^2.

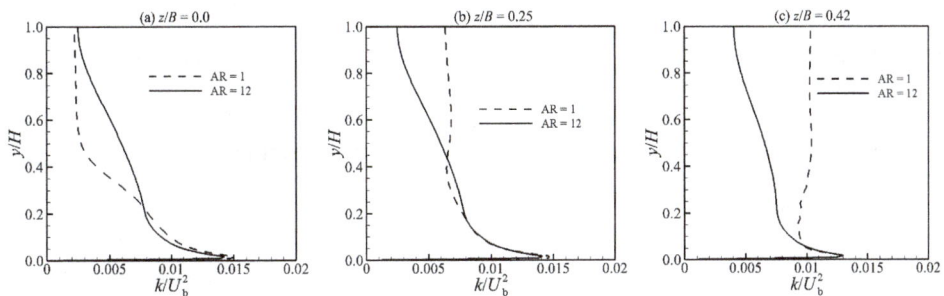

Figure 6:   Variation of the normalized turbulent kinetic energy k/U_b^2 profiles of channel flow in the fully-developed region with the flow depth y/H for AR = 1 and 12 extracted at: (a) $z/B = 0$; (b) $z/B = 0.25$; and (c) $z/B = 0.42$.

6 CONCLUSIONS

From the above, the following conclusions can be drawn:

- A pair of counter-rotating mean recirculation zones is formed at the bottom edges of a straight open channel flow; one near the sidewall and one near the bed. The recirculation zone that reside near the bed is dominant and energetic and almost spans the entire flow depth for AR = 12, while the one that reside near the sidewall appears weaker, more confined, and its size is much smaller. Contrary to this behavior, the near-bed recirculation zone becomes more confined and relatively smaller for the AR = 1 case compared to the AR = 12 case, while the ones that reside near the sidewalls become much more energetic and larger in size, and their influence extend up to the free surface.
- The distribution of the turbulent kinetic energy k indicates that the maximum value occurs in the layer adjacent to the solid walls for both aspect ratios, and decreases gradually away from the solid walls towards the core region towards the free surface. However, the magnitude of k appears to increase with a much higher rate as the sidewalls are approached for AR = 1 compared to the AR = 12 case, at corresponding z/B. It can be inferred that the decrease of the AR causes an increase of the components of turbulent velocity fluctuations.

ACKNOWLEDGEMENTS

This research was made possible by the facilities of the Shared Hierarchical Academic Computing Network (SHARCNET: www.sharcnet.ca) and Compute/Calcul Canada.

REFERENCES

[1]   Bradshaw, P., Turbulent secondary flows. *Annual Review of Fluid Mechanics*, **19**(1), pp. 53–74, 1987.
[2]   Einstein, H. & Li, H., Secondary currents in straight channels. *EOS, Transactions American Geophysical Union*, **39**(6), pp. 1085–1088, 1958.
[3]   Nezu, I. & Rodi, W., Experimental study on secondary currents in open-channel flows. *Proceedings of 21st IAHR Congress*, Melbourne, pp. 115–119, 1985.
[4]   Nezu, I., Nakagawa, H. & Tominaga, A., Secondary currents in a straight channel flow and the relation to its aspect ratio. *In Turbulent Shear Flows*, **4**, pp. 246–260, 1985.
[5]   Nezu, I. & Nakagawa, H., *Turbulence in Open Channel Flows*, Balkema: Netherlands, 1993.
[6]   Tominaga, A., Nezu, I., Ezaki, K. & Nakagawa, H., Three-dimensional turbulent structure in straight open channel flows. *Journal of Hydraulic Research*, **27**(1), pp. 149–173, 1989.
[7]   Fenton, J.D., The Cnoidal theory of water waves. *Developments in Offshore Engineering*, pp. 55–100, 1999.
[8]   Nasif, G., Balachandar, R. & Barron, R.M., Supercritical flow characteristics in smooth open channels with different aspect ratios. *Physics of Fluids*, **32**(10), p. 105102, 2020.
[9]   Siemens PLM, *Global, STAR-CCM+ 13.06.012: User Manual*, 2018.
[10]   Malalasekera, W. & Versteeg, H.K., *An Introduction to Computational Fluid Dynamics: The Finite Volume Method*, 2nd edn, Pearson Education Ltd.: Harlow, UK, 2007.
[11]   Menter, F.R. & Kuntz, M., Adaptation of eddy-viscosity turbulence models to unsteady separated flow behind vehicles. *The Aerodynamics of Heavy Vehicles: Trucks, Buses, and Trains*, Springer: Berlin, Heidelberg, pp. 339–352, 2004.

[12] Travin, A., Shur, M., Strelets, M.M. & Spalart, P.R., Physical and numerical upgrades in the detached-eddy simulation of complex turbulent flows. *Advances in LES of Complex Flows*, pp. 239–254, 2002.

[13] Spalart, P.R., Deck, S., Shur, M.L., Squires, K.D., Strelets, M.K. & Travin, A., A new version of detached-eddy simulation, resistant to ambiguous grid densities. *Theoretical and Computational Fluid Dynamics*, **20**(3), p. 181, 2006.

[14] Wilcox, D.C., Simulation of transition with a two-equation turbulence model. *AIAA Journal*, **32**(2), pp. 247–255, 1994.

[15] Wilcox, D.C., *Turbulence Modelling for CFD*, vol. 2, DCW Industries: La Canada, CA, pp. 103–217, 1998.

[16] Hirt, C.W. & Nichols, B.D., Volume of fluid (VOF) method for the dynamics of free boundaries. *Journal of Computational Physics*, **39**(1), pp. 201–225, 1981.

[17] Leonard, B.P., The ULTIMATE conservative difference scheme applied to unsteady one-dimensional advection. *Computer Methods in Applied Mechanics and Engineering*, **88**(1), pp. 17–74, 1991.

[18] Ubbink, O. & Issa, R.I., Method for capturing sharp fluid interfaces on arbitrary meshes. *Journal of Computational Physics*, **153**, pp. 26–50, 1999.

[19] Muzaferija, S., A two-fluid Navier–Stokes solver to simulate water entry. *Proceedings of 22nd Symposium on Naval Architecture*, pp. 638–651, 1999.

[20] Heidari, M., Balachandar, R., Roussinova, V. & Barron, R.M., Characteristics of flow past a slender, emergent cylinder in shallow open channels. *Physics of Fluids*, **29**(6), p. 065111, 2017.

[21] Nasif, G., Balachandar, R. & Barron, R.M., Influence of bed proximity on the three-dimensional characteristics of the wake of a sharp-edged bluff body. *Physics of Fluids*, **31**(2), p. 025116, 2019.

[22] Nasif, G., Balachandar, R. & Barron, R.M., Effect of gap on the flow characteristics in the wake of a bluff body near a wall. *International Journal of Computational Methods and Experimental Measurements*, **7**(4), pp. 305–315, 2019.

[23] Kirgoz, M.S. & Ardiclioglu, M., Velocity profiles of developing and developed open channel flow. *Journal of Hydraulic Engineering*, **123**(12), pp. 1099–1105, 1997.

[24] Auel, C., Albayrak, I. & Boes, M., Turbulence characteristics in supercritical open channel flows: Effects of Froude number and aspect ratio. *Journal of Hydraulic Engineering*, **140**(4), p. 04014004-1-15, 2014.

[25] Clauser, F., Turbulent boundary layers in adverse pressure gradient. *Journal of the Aerospace Sciences*, **21**, pp. 91–108, 1954.

[26] Balachandar, R. & Bhuiyan, F., Higher-order moments of velocity fluctuations in an open channel flow with large bottom roughness. *Journal of Hydraulic Engineering*, **133**, pp. 77–87, 2007.

SECTION 2
NANO AND MICRO FLUIDS

CURVATURE AND TEMPERATURE EFFECT ON *n*-DECANE TRANSPORT IN NARROW CARBON NANOTUBES

ZHONGLIANG CHEN[1]*, XIAOHU DONG[1] & ZHANGXIN CHEN[2]
[1]State Key Laboratory of Petroleum Resources and Prospecting, China University of Petroleum, China
[2]Department of Chemical and Petroleum Engineering, University of Calgary, Canada

ABSTRACT
Carbon nanotubes (CNTs) are excellent materials for advanced functional nano-elements. They have superior mechanical, electronic, and chemical properties and are widely used in many fields, such as nanomechanics, advanced electronics, biotechnology, and energy. Alkanes transport through CNTs has received widespread attention. This study aims to propose a systematic method to study the coupling effect of curvature and temperature on the *n*-decane transport through narrow CNTs. The OPLS (optimized potentials for liquid simulations) model and Lennard-Jones potential are used to describe the intermolecular/intramolecular interactions in a typical *n*-decane/CNT system. All molecular dynamics (MD) simulations are conducted in the NVT ensemble to show the dynamic of *n*-decane molecules in 1.08, 1.36, and 2.71 nm-diameter single-walled armchair CNTs. The Green–Kubo and Stokes–Einstein expression are combined with the MD simulations to calculate the *n*-decane/CNT friction coefficient, the *n*-decane axial self-diffusion coefficient, and viscosity in CNTs. The results show that increased curvature causes the *n*-decane/CNT friction coefficient to decline rapidly. However, the changes in the axial self-diffusion coefficient and viscosity are non-monotonic. On the contrary, the effect of increasing temperature is just the opposite; that is, for individual CNTs, the axial self-diffusion coefficient generally increases, and the viscosity decreases, but the friction coefficient fluctuates. We also find that the non-monotonic change between the curvature and the axial self-diffusion coefficient is substantially temperature-independent. An increase in temperature has a positive effect on the axial diffusion of *n*-decane molecules. However, when the curvature of carbon nanotubes is too large (the 1.08 nm-diameter CNT), there is no way to sustain this positive effect. It is worth emphasizing that even with high temperatures, a CNT with a more significant curvature does not mean that *n*-decane is more difficult to transport through.
Keywords: carbon nanotube, curvature, n-decane, friction coefficient, diffusion.

1 INTRODUCTION
Nanofluidics is a study for the behaviour, manipulation and control of fluids that are confined to nanostructures. This process has experienced considerable growth in recent years [1]. Due to the characteristic physical scaling lengths of fluid closely coincide with the dimensions of the nanostructure itself, new solutions and properties can be obtained from the scales where the behavior of matter departs from conventional expectations [2], [3]. The spatial structure of the forces acting on the nanoscale must be fully taken into account to understand how fluids behave [4].

CNTs are excellent materials for advanced functional nano-elements. They have superior mechanical, electronic and chemical properties, and are widely used in many fields [5], [6], such as nanomechanics, advanced electronics, biotechnology and energy. MD simulations, acting as a bridge between microscopic length and time scales and the macroscopic world of laboratory [7], allow us to focus on the dynamical properties of a typical fluid/CNT system,

---

* ORCID: *https://orcid.org/0000-0002-8881-2294*

such as transport coefficients, time-dependent responses to perturbations and rheological properties.

As an intrinsically interfacial property, the inverse hyperbolic sine relationship [8] shows that the friction dominates the transport behaviour of water at the wall of CNT. Even increasing the interactions between water and CNT by modifying the membranes, it still produces non-uniform nanofluid flow with lower friction than that consistent with the Navier–Stokes equations [9]. Once we calculate the activation energy required for the transportation process, we can interpret the transport behaviour and predict the friction coefficient [10] between water and CNT. However, it is slightly different from the results [11] determined using the Green-Kubo relationship of the liquid-solid friction coefficient. There is also a significant flow enhancement for decane through nanoscale CNTs [12], which can be explained by the lower friction coefficient at carbon interface. The flow rate of decane is four to five orders of magnitude faster than conventional fluid flow would predict through 7 nm-diameter CNT [13], and the observed slip length (3.4 μm) are much longer than the pore radius (3.5 nm) that is consistent with a nearly frictionless interface.

The values of water self-diffusion coefficient vary widely in different CNTs, depending on the density, temperature, and confinement [14]. However, for a set of narrow CNTs with the same cross-sectional area but different cross-sectional shapes, the mobilities of water can also differ considerably [15]. Furthermore, it is possible to separate the effect of the CNT surface and the effect of the cross-sectional shape of the confinement [16], and the diffusion mechanisms of ballistic, Fickian and single-file types can be determined by analyzing the style of the time evolution of the mean squared displacement [17]. For n-decane molecules, the anomalous positive peaks are observed in the velocity autocorrelation function perpendicular to the axis of CNT, which can be explained by the oscillating motion of the molecules trapped in the effective potential well produced by the wall of CNT [18], and the diffusion of molecules in this area is different from that in the central area of CNT.

Viscosity is a physical property of a fluid that opposes the relative motion between two liquid layers, or in a fluid that moves at different velocities [19]. For a larger set of CNTs, the viscosity of the confined water increases with the increase of temperature and CNT diameter, while the size-dependent trend of viscosity is almost independent of temperature [20]. When we consider a single file or a single layer of molecules transport through a narrow CNT, the calculation of viscosity based on the Eyring theory of reaction rates seems not suitable for this case [21]. However, the equilibrium molecular dynamics simulation does not require other adjustments that the nonequilibrium method usually needs, and has shown advantages in solving such problems [22]. Although Einstein's model [23] does not include the structural parameters of the flow channel [24], the diffusion coefficient in this model is very sensitive to the curvature [25] and even flexibility [26] of the CNTs.

The main goal of this paper is to propose a systematic method to study the coupling effect of curvature and temperature on the n-decane transport through narrow CNTs. The focus is two-fold. First, the combination of theory and MD simulations is emphasized, including the theoretical background of the friction coefficient and transport coefficient based on Green–Kubo expression and the Stokes–Einstein relationship, as well as the process of model building and the selection of essential parameters. Second, focus on the core of the subject, which is to use MD simulation to understand how curvature and temperature affect the transport of n-decane in narrow CNTs. Although some notions are at the historical foundation of the subject, new systematic practice and exciting results of MD simulations have recently come to light.

2 THEORY AND MD SIMULATION

2.1 Liquid-solid friction coefficient

Liquid–solid friction coefficient affects the fluid flow and permeability of porous media, which is often used as an essential parameter in models of transport properties through micro- and nano-channels. A geometry model of the Brownian motion of a cylindrical wall (CNT) in contact with a liquid (n-decane) is introduced and a non-Marconian Langevin equation (eqn (1)) of motion of the fluctuating wall [27], including the liquid-solid friction coefficient, λ, is established.

$$m\frac{dU}{dt} = -\lambda A v_s\left(t\right) + \delta F\left(t\right).$$ (1)

In eqn (1), A is the lateral surface area, $\delta F(t)$ is the lateral fluctuating force, m and $U(t)$ are the mass and velocity of the wall, respectively. The hydrodynamic slip velocity, $v_s(t)$, which describes the velocity discontinuity between the wall and the liquid, can be defined as [27],

$$v_s\left(t\right) = \int_{-\infty}^{+\infty} dt' K\left(t-t'\right) U\left(t'\right),$$ (2)

where, $K(t)$ is the memory kernel. The introduction of the memory kernel with vanishing time integral allows us to have a better acquaintance with the real slip velocity in the linear response regime. We use the Laplace transform of the velocity autocorrelation function $C(t)$ = $<U(t)U(0)>$ and the sum rules of the memory kernel to calculate the Laplace transform of the autocorrelation function of the axial component of the force acting on the lateral surface [27]. The Green–Kubo relation for the friction coefficient is found by integrating the force autocorrelation function over time [27],

$$\lambda = \frac{1}{Ak_BT}\int_0^\infty dt\left\langle F_z\left(t\right)\cdot F_z\left(0\right)\right\rangle,$$ (3)

where, k_B is the Boltzmann constant and T is the temperature.

2.2 Diffusion coefficient and viscosity

The axial self-diffusion coefficient [28], [29] (in one dimension), D_z, valid at long times, can be written as,

$$D_z = \lim_{t\to\infty}\frac{d}{dt}\frac{1}{2}\left\langle\left|z\left(t\right)-z\left(0\right)\right|^2\right\rangle = \lim_{t\to\infty}\frac{d}{dt}\frac{1}{2}\left\langle\frac{1}{N}\sum_{i=1}^{N}\left|z_i\left(t\right)-z_i\left(0\right)\right|^2\right\rangle.$$ (4)

For each of the N atoms in the simulation, the centre-of-mass axial position is z_i, $<|z(t)-z(0)|^2>$ is the mean squared displacement (MSD) [30], [31].

And the corresponding shear viscosity, η, is found to be proportional to the mean squared x displacement of the centre of y momentum [30], [32], [33].

$$\eta = \lim_{t\to\infty}\frac{d}{dt}\frac{1}{2}\frac{V}{k_BT}\left\langle\left[Q_{xy}\left(t\right)-Q_{xy}\left(0\right)\right]^2\right\rangle,$$ (5)

where V is a volume of the particle system and the dynamical variable, Q_{xy}, can be defined as,

$$Q_{xy} = \frac{1}{V}\sum_{i=1}^{N} x_i m_i v_{iy},$$ (6)

where m_i, x_i and v_{iy} are the mass, the centre-of-mass radial position and velocity for each of the N atoms, respectively.

Obviously, given the existing axial self-diffusion coefficient, D_z, we also can predict the viscosity by evoking the Stokes–Einstein, SE, relationship [28], i.e. the theory of Brownian motion for a particle slightly more massive than or approximately equal to the solvent molecules [34], [35], of diameter $2r$, immersed in a liquid leads to a relationship [36] between the shear viscosity, η, of the host liquid and the axial self-diffusion coefficient, D_z, of the particle,

$$D_z = \frac{k_B T}{4\pi\eta r},$$ (7)

where, the effective diameter [34] for an n-decane molecule is 4.2 Å.

2.3 Simulation parameters and model

n-Decane interactions are modelled using the OPLS (optimized potentials for liquid simulations) model [37], CH_n groups are treated as united atoms centered on the carbon [38], [39], i.e. every methyl (CH_3) or methylene (CH_2) group is modelled as a single interaction site [40]. The equilibrium value of the distance between neighboring sites [39], [41] (the bond length) is 1.53 Å, the equilibrium value of the angle between two connected bonds [39], [42] (the bond angle) is 112°, and the harmonic force constants for bonds [41], [43] and angles [11], [44] are 900 kcal/mol/Å and 124.2 kcal/mol/rad², respectively. The Fourier series (eqn (8)) can describe the rotational potential energy [37] and let the dihedral angle C-C-C-C vary over a sufficiently broad range [39].

$$V(\phi) = \frac{1}{2}\times 1.411\times(1+\cos\phi) - \frac{1}{2}\times 0.271\times(1-\cos 2\phi) + \frac{1}{2}\times 3.145\times(1+\cos 3\phi).$$ (8)

The Lennard–Jones potential suffices to describe the non-bonded interactions between the united atoms from different molecules and within a molecule [40] (if two atoms are more than four atoms apart), and the corresponding optimized Lennard–Jones parameters are listed in Table 1. A standard mixing rule, $\varepsilon_{ij} = \sqrt{\varepsilon_i \varepsilon_j}$ and $\sigma_{ij} = \frac{1}{2}\left(\sigma_i + \sigma_j\right)$, is used to create pair coefficients for interactions between different (united) atoms.

Table 1: Optimized Lennard–Jones parameters for n-decane (XX) and n-decane-CNT (XC) interaction.

X	σ_{XX} (Å)	ε_{XX} (kcal/mol)	σ_{XC} (Å)	ε_{XC} (kcal/mol)
CH_3	3.905	0.175	3.828	0.136
CH_2	3.905	0.118	3.828	0.111

The nanotube-builder for VMD [45] is used to generate 1.08, 1.36, and 2.71 nm-diameter single-walled armchair CNTs with smooth tips, all of which are 10 nanometers long. We do not plan on defining bonded interactions between carbon atoms. A snapshot [45] from a typical *n*-decane/CNT system is presented in Fig. 1, and the chirality vector for each CNT is listed.

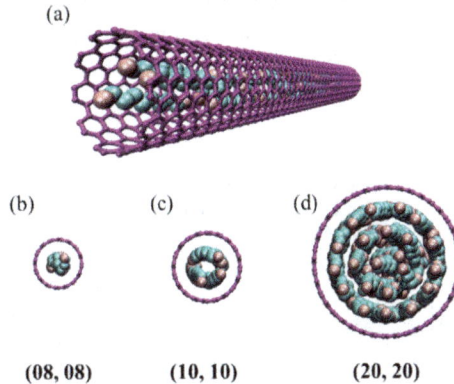

(a)

(b)   (c)   (d)

(08, 08)   **(10, 10)**   **(20, 20)**

Figure 1:   (a) A snapshot from a typical *n*-decane/CNT system. The diameters of the single-walled armchair CNTs are 1.08 nm (b), 1.36 nm (c) and 2.71 nm (d) respectively.

n-Decane-carbon friction and *n*-decane diffusion coefficient can be predicted by using a Green–Kubo relation in an equilibrium (no net flow) simulation [11], [27], but until then, an additional equilibrium simulation [11] is needed to push *n*-decane molecules inside the tube using a piston-like mechanism. The pressure applied to the piston is one atmosphere [39], which ensures that there are a suitable number of *n*-decane molecules in CNTs of different sizes. After reaching equilibrium, the pistons and *n*-decane reservoirs on both sides of the CNT are removed.

All simulations are performed in the NVT ensemble (constant mass, volume, and temperature) with a temperature maintained at a specific value using a Nosé–Hoover thermostat [46], time integration is performed on Nosé–Hoover style non-Hamiltonian equations of motion which are designed to generate positions and velocities sampled from the canonical ensemble. The temperature is relaxed in a time span of 0.2 ps [39] in anticipation of a balance between milder temperature fluctuations and less equilibration time. In order to guarantee the quality of the equilibrium simulations, the temperature is changed from 300 K to 360 K by 3 K steps, using the final configuration from the previous temperature as initial state [47].

The total force between the CH_n groups and the carbon atoms is measured every 2 fs [39] to calculate the autocorrelation function of the axial component of the force acting on the wall of the CNT (eqn (3)). Once the linear state in MSD (eqn (4)) is reached, the axial self-diffusion coefficient can be better estimated at an early stage rather than at later correlation times [48]. Successive time origins [49] are set to produce the Green-Kubo curves and the MSD curves, adequate statistics and the standard error of the estimate are used [47], [49] for determination of the *n*-decane/CNT friction coefficient and the *n*-decane axial self-diffusion coefficient, error bars in the simulations are roughly similar or smaller than the symbols in the following figures.

3 RESULTS AND DISCUSSION

3.1 *n*-Decane-carbon friction coefficient

As shown in Fig. 2, our MD simulation results for the *n*-decane/CNT friction coefficient are consistent with the equilibrium and flow MD simulation results in the published paper [39]. The results of the water/CNT friction coefficient [39] are also shown for comparison. Interestingly, although the structure of the *n*-decane molecule is larger and more complicated than that of water molecules, the *n*-decane/CNT friction coefficient is smaller. And they all decrease with the increase in curvature.

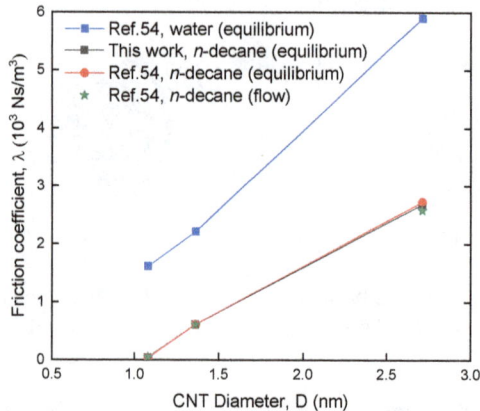

Figure 2:   Variati on in water/CNT and *n*-decane/CNT friction coefficient with CNT diameter *D*. We compared the MD results of the *n*-decane/CNT friction coefficient (eqn (3)) with the published results [39], including the equilibrium and nonequilibrium (flow) MD simulations. The water/CNT friction coefficient [39] is also shown as a reference. All of these simulations have a temperature of 300 K.

We also find that this trend remains the same, even when affected by temperature-related changes (Fig. 3(a)). But if we look more closely, the plot actually thickens (Fig. 3(b)), i.e. for individual CNTs, the change of the curve is non-monotonous due to the coupling effect of curvature and temperature.

The friction coefficient fluctuates more dramatically in smaller CNTs. For the 2.71 nm-diameter single-walled armchair CNT, raising the temperature generally reduce the friction coefficient, which is quite the opposite for the smallest CNT, and the situation is somewhere in between in the 1.36 nm-diameter CNT.

3.2 *n*-Decane diffusion coefficient

We can think of *n*-decane as an ellipsoid [50], and if the rotation of the uniaxial anisotropic particle is prohibited, its diffusion will mainly occur in the direction parallel to its long axis [51], but even if the rotation is permitted, according to the results [52] obtained using the diffusion map approach to observe such molecules with high aspect ratios in aqueous

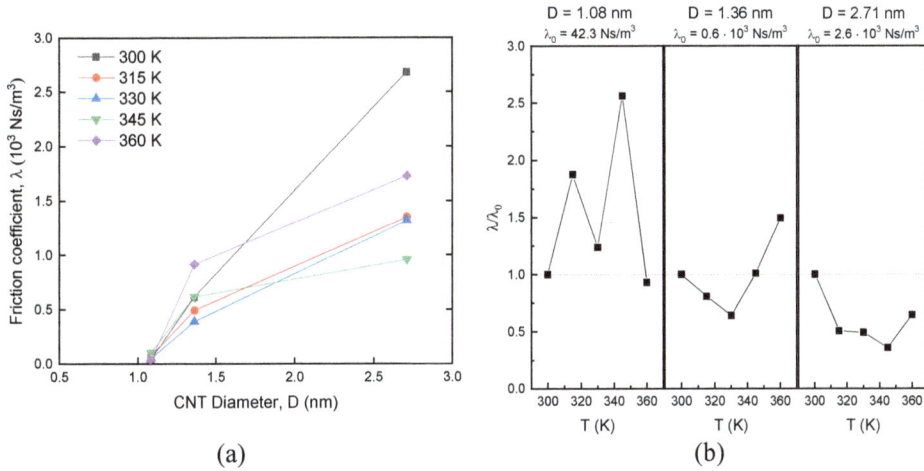

Figure 3:   Curvature and temperature effect on *n*-decane/CNT friction coefficient, as predicted from equilibrium MD simulation using eqn (3). (a) Friction coefficient versus CNT diameter; and (b) Friction coefficient versus temperature, λ_0 is the *n*-decane/CNT friction coefficient at 300 K.

solution, as one of the ordered principal moments, ζ_1 describes the extent of the molecule along its longest axis and contributes the most to the radius of gyration.

If we go the extra mile and go from aqueous solution to CNTs (Fig. 4), we will find that *n*-decane molecules tend to line along the axial direction in CNTs, such a structural arrangement (i.e. the orientational ordering) would lead to a relatively small displacement in the radial direction [11], [53].

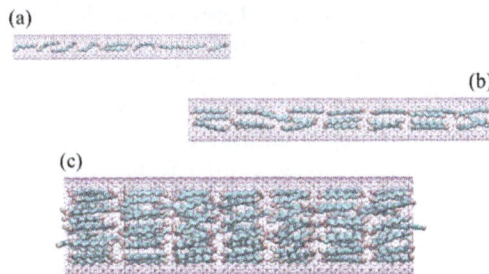

Figure 4:   The structural arrangement of *n*-decane molecules in CNTs. The diameters of the single-walled armchair CNTs are 1.08 nm (a), 1.36 nm (b) and 2.71 nm (c) respectively.

Different from the calculation results of the friction coefficient, the axial self-diffusion coefficient of *n*-decane does not show a monotonic change with the decreasing diameter of CNTs (Fig. 5(a)). We believe that this non-monotonicity is the result of the tripartite game among the curvature, the depletion area [11], [54] at the *n*-decane/CNT interface and the central area.

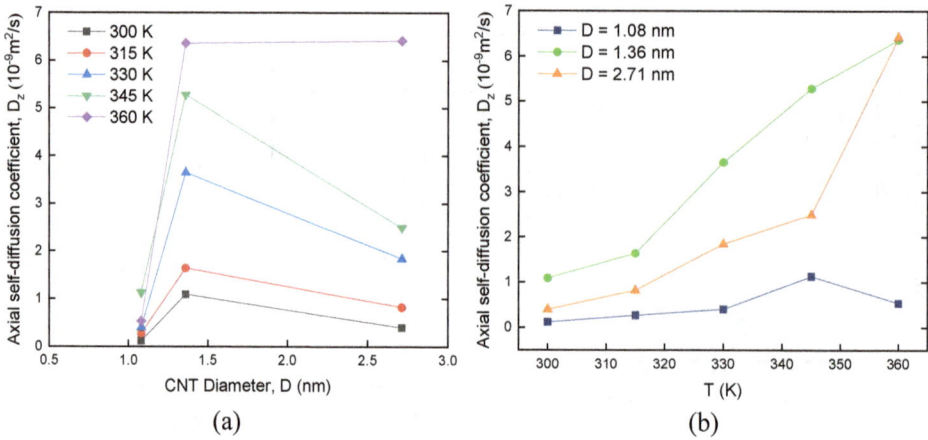

Figure 5: Curvature and temperature effect on axial self-diffusion coefficient of n-decane, as predicted from equilibrium MD simulation using eqn (4). (a) Axial self-diffusion coefficient versus CNT diameter; and (b) Axial self-diffusion coefficient versus temperature.

For the 1.36 nm-diameter single-walled armchair CNT, a depletion area at the n-decane/CNT interface plays a dominant role in affecting the axial diffusion of n-decane molecules, which can be shown by the complexity of the spatial variations (from the centre to the nanotube wall) [55] in the axial self-diffusion coefficient, i.e. the more significant the axial diffusion in the area closer to the wall of CNTs.

For the 1.08 nm-diameter CNT, excessive curvature is primarily responsible for the reduction of the axial self-diffusion coefficient of n-decane molecules. However, for the 2.71 nm-diameter CNT, the effects from curvature and depletion area are more or less weakened, and the bulk diffusion of the central area becomes evident.

Further shifting the focus to changes in temperature, we find that this non-monotonic change between the curvature and the axial self-diffusion coefficient is substantially temperature-independent (Fig. 5(a)). For individual CNTs, an increase in temperature generally has a positive effect on the axial diffusion of molecules (Fig. 5(b)). However, when the curvature of carbon nanotubes is too large (the 1.08 nm-diameter CNT), there is no way to sustain this positive effect.

3.3 *n*-Decane viscosity

Whether from a theoretical [39], [56], [57] or experimental [58] perspective, calculating the viscosity of the liquid in such small-sized CNTs is full of challenges. The sliding of the layers required for shear flow is severely negatively affected by the CNT confinement in the radial direction [56], and the thermal fluctuations of the shear stress in the equilibrium state is therefore significantly different from that in the bulk system [57].

For a typical water/CNT system, when the diameter of the CNT is less than 3 nm, the elongated tubular structure seriously affects the arrangement and movement of water molecules, resulting in the radial viscosity [56] of water being small enough to be ignored.

It is clear that the orientational ordering of n-decane molecules is more likely to be parallel to the axis of CNTs. Especially when the curvature is very high, the molecules desperately hope to avoid bending caused by the curvature of the wall in the radial direction. So, we use the Stokes–Einstein relationship to predict the n-decane viscosity, which is an attempt from the perspective of the axial self-diffusion coefficient.

The relationship between n-decane viscosity and curvature of CNTs at temperatures between 300 K and 360 K is presented in Fig. 6, which is diametrically opposite to the change of the axial self-diffusion coefficient of n-decane. The viscosity of n-decane does not decrease monotonically with the decrease of pore size. Comparing this trend with several MD results for water viscosity, some of which show similar changes [39], [57], while others are different [28], [56].

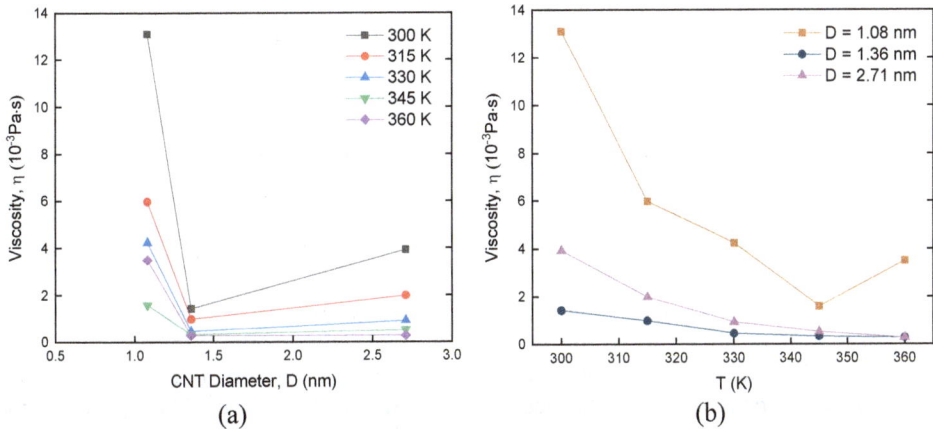

Figure 6:   Curvature and temperature effect on n-decane viscosity, as predicted from equilibrium MD simulation using eqn (7). (a) Viscosity versus CNT diameter and (b) Viscosity versus temperature.

The raised temperature cannot change the non-monotonic variation between the curvature and the viscosity (Fig. 6(a)). And for individual CNTs, the relationship between temperature and n-decane viscosity is relatively simple (Fig. 6(b)), i.e. in general, increasing the temperature will cause a decrease in n-decane viscosity. It is important to note that for the 1.08 nm-diameter CNT, this trend will change when the effect of temperature dominates.

4 CONCLUSIONS

More significant curvature and temperature changes can seriously affect the diffusion coefficient and viscosity of n-decane, as well as the friction coefficient between it and the wall of CNT. The OPLS model and Lennard–Jones potential is used to describe the intermolecular/intramolecular interactions in a typical n-decane/CNT system. All MD simulations are conducted in the NVT ensemble to show the dynamic of n-decane molecules in 1.08, 1.36, and 2.71 nm-diameter single-walled armchair CNTs.

Increased curvature causes the n-decane/CNT friction coefficient to decline rapidly; however, the changes in the axial self-diffusion coefficient and viscosity are non-monotonic. On the contrary, the effect of increasing temperature is just the opposite, that is, for individual CNTs, the axial self-diffusion coefficient generally increases, and the viscosity decreases, but the friction coefficient fluctuates.

The change in temperature makes the tripartite game among the curvature, the depletion area at the n-decane/CNT interface and the central area a little more complicated, especially for the 1.08 nm-diameter CNT. Therefore, it is worth emphasizing that even with high temperatures, a CNT with a more significant curvature does not mean that n-decane is more difficult to transport through.

ACKNOWLEDGEMENTS

This work was financially supported by the Beijing Natural Science Foundation (Grant No. 3212020) and the National Natural Science Foundation of China (Grant No. 52004303).

REFERENCES

[1] Schoch, R.B., Han, J. & Renaud, P., Transport phenomena in nanofluidics. *Rev. Mod. Phys*, **80**, p. 839, 2008.
[2] Bocquet, L. & Charlaix, E., Nanofluidics from bulk to interfaces. *Chem. Soc. Rev*, **39**, p. 1073, 2010.
[3] Dong, X., Liu, H., Guo, W., Hou, J., Chen, Z. & Wu, K., Study of the confined behavior of hydrocarbons in organic nanopores by the potential theory. *Fluid Phase Equilib.*, **429**, p. 214, 2016.
[4] Bocquet, L. & Tabeling, P., Physics and technological aspects of nanofluidics. *Lab Chip*, **14**, p. 3143, 2014.
[5] Hirai, Y. et al., Molecular dynamics studies on mechanical properties of carbon nano tubes with pinhole defects. *Jpn. J. Appl. Phys.*, **42**, p. 4120, 2003.
[6] Dong, X., Liu, H., Hou, J., Wu, K. & Chen, Z., Phase equilibria of confined fluids in nanopores of tight and shale rocks considering the effect of capillary pressure and adsorption film. *Ind. Eng. Chem. Res.*, **55**, p. 798, 2016.
[7] Allen, M.P. et al., Introduction to molecular dynamics simulation. *Comput. Soft Matter from Synth. Polym. to Proteins*, **23**, p. 1, 2004.
[8] Ma, M.D., Shen, L., Sheridan, J., Liu, J.Z., Chen, C. & Zheng, Q., Friction of water slipping in carbon nanotubes. *Phys. Rev. E*, **83**, p. 36316, 2011.
[9] Wang, L., Dumont, R.S. & Dickson, J.M., Nonequilibrium molecular dynamics simulation of pressure-driven water transport through modified CNT membranes. *J. Chem. Phys.*, **138**, p. 124701, 2013.
[10] Babu, J.S. & Sathian, S.P., Combining molecular dynamics simulation and transition state theory to evaluate solid-liquid interfacial friction in carbon nanotube membranes. *Phys. Rev. E*, **85**, p. 51205, 2012.
[11] Falk, K., Sedlmeier, F., Joly, L., Netz, R.R. & Bocquet, L., Ultralow liquid/solid friction in carbon nanotubes: Comprehensive theory for alcohols, alkanes, OMCTS, and water. *Langmuir*, **28**, p. 14261, 2012.
[12] Whitby, M., Cagnon, L., Thanou, M. & Quirke, N., Enhanced fluid flow through nanoscale carbon pipes. *Nano Lett.*, **8**, p. 2632, 2008.
[13] Majumder, M., Chopra, N., Andrews, R. & Hinds, B.J., Enhanced flow in carbon nanotubes. *Nature*, **438**, p. 44, 2005.
[14] Tsimpanogiannis, I.N., Moultos, O.A., Franco, L.F.M., de M, M.B. Spera, M. Erdős, & Economou I.G., Self-diffusion coefficient of bulk and confined water: A critical review of classical molecular simulation studies. *Mol. Simul.*, **45**, p. 425, 2019.
[15] Nie, G.X., Wang, Y. & Huang, J.P., Shape effect of nanochannels on water mobility. *Front. Phys.*, **11**, p. 114702, 2016.
[16] Zheng, Y., Ye, H., Zhang, Z. & Zhang, H., Water diffusion inside carbon nanotubes: Mutual effects of surface and confinement. *Phys. Chem. Chem. Phys.*, **14**, p. 964, 2012.

[17] Striolo, A., The mechanism of water diffusion in narrow carbon nanotubes. *Nano Lett.*, **6**, p. 633, 2006.

[18] Zhang, F., Molecular dynamics studies of chainlike molecules confined in a carbon nanotube. *J. Chem. Phys.*, **111**, p. 9082, 1999.

[19] Jabbari, F., Rajabpour, A. & Saedodin, S., Thermal conductivity and viscosity of nanofluids: A review of recent molecular dynamics studies. *Chem. Eng. Sci.*, **174**, p. 67, 2017.

[20] Ye, H., Zhang, H., Zhang, Z. & Zheng, Y., Size and temperature effects on the viscosity of water inside carbon nanotubes. *Nanoscale Res. Lett.*, **6**, p. 87, 2011.

[21] Babu, J.S. & Sathian, S.P., The role of activation energy and reduced viscosity on the enhancement of water flow through carbon nanotubes. *J. Chem. Phys.*, **134**, p. 194509, 2011.

[22] Vakili-Nezhaad, G., Al-Wadhahi, M., Gujrathi, A.M., Al-Maamari, R. & Mohammadi, M., Effect of temperature and diameter of narrow single-walled carbon nanotubes on the viscosity of nanofluid: A molecular dynamics study. *Fluid Phase Equilib.*, **434**, p. 193, 2017.

[23] Einstein, A., *Eine Neue Bestimmung Der Moleküldimensionen*, ETH Zurich, 1905.

[24] Kumaresan, V. & Velraj, R., Experimental investigation of the thermo-physical properties of water–ethylene glycol mixture based CNT nanofluids. *Thermochim. Acta*, **545**, p. 180, 2012.

[25] Sahu, P. & Ali, S.M., Curious characteristics of polar and nonpolar molecules confined within carbon nanotubes (CNT) of varied diameter: Insights from molecular dynamics simulation. *J. Chem. Eng. Data*, **62**, p. 2307, 2017.

[26] Jakobtorweihen, S., Lowe, C.P., Keil, F.J. & Smit, B., A novel algorithm to model the influence of host lattice flexibility in molecular dynamics simulations: Loading dependence of self-diffusion in carbon nanotubes. *J. Chem. Phys.*, **124**, p. 154706, 2006.

[27] Bocquet, L. & Barrat, J.L., On the green-Kubo relationship for the liquid-solid friction coefficient. *J. Chem. Phys.*, **139**, p. 44704, 2013.

[28] Thomas, J.A., McGaughey, A.J.H. & Kuter-Arnebeck, O., Pressure-driven water flow through carbon nanotubes: Insights from molecular dynamics simulation. *Int. J. Therm. Sci.*, **49**, p. 281, 2010.

[29] Mashl, R.J., Joseph, S., Aluru, N.R. & Jakobsson, E., Anomalously immobilized water: A new water phase induced by confinement in nanotubes. *Nano Lett.*, **3**, p. 589, 2003.

[30] Allen, M.P. & Tildesley, D.J., *Computer Simulation of Liquids*, Oxford University Press, 2017.

[31] Zwanzig, R.W., *Nonequilibrium Statistical Mechanics*, Oxford University Press, 2001.

[32] Helfand, E., Transport coefficients from dissipation in a canonical ensemble. *Phys. Rev.*, **119**, p. 1, 1960.

[33] Cheng, A. & Merz, K.M., The pressure and pressure tensor for macromolecular systems. *J. Phys. Chem.*, **100**, p. 905, 1996.

[34] Falk, K., Coasne, B., Pellenq, R., Ulm, F.J. & Bocquet, L., Subcontinuum mass transport of condensed hydrocarbons in nanoporous media. *Nat. Commun.*, **6**, p. 6949, 2015.

[35] Li, J.C.M. & Chang, P., Self-diffusion coefficient and viscosity in liquids. *J. Chem. Phys.*, **23**, p. 518, 1955.

[36] Heyes, D.M., *The Liquid State: Applications of Molecular Simulations*, Chichester, 1998.

[37] Jorgensen, W.L., Madura, J.D. & Swenson, C.J., Optimized intermolecular potential functions for liquid hydrocarbons. *J. Am. Chem. Soc.*, **106**, p. 6638, 1984.

[38] Jorgensen, W.L. & Tirado-Rives, J., The OPLS [Optimized potentials for liquid simulations] potential functions for proteins, energy minimizations for crystals of cyclic peptides and crambin. *J. Am. Chem. Soc.*, **110**, p. 1657, 1988.

[39] Falk, K., *The Molecular Origin of Fast Fluid Transport in Carbon Nanotubes: Theoretical and Molecular Dynamics Study of Liqui/Solid Friction in Graphitic Nanopores*, Université Claude Bernard – Lyon I, 2011.

[40] Nicolas, J.P. & Smit, B., Molecular dynamics simulations of the surface tension of N-Hexane, n-Decane and n-Hexadecane. *Mol. Phys.*, **100**, p. 2471, 2002.

[41] Mundy, C.J., Siepmann, J.I. & Klein, M.L., Calculation of the shear viscosity of decane using a reversible multiple time-step algorithm. *J. Chem. Phys.*, **102**, p. 3376, 1995.

[42] Smit, B., Karaborni, S. & Siepmann, J.I., Computer simulations of vapor–liquid phase equilibria of N-alkanes. *J. Chem. Phys.*, **102**, p. 2126, 1995.

[43] Cui, S.T., Cummings, P.T. & Cochran, H.D., Multiple time step nonequilibrium molecular dynamics simulation of the rheological properties of liquid N-decane. *J. Chem. Phys.*, **104**, p. 255, 1996.

[44] Supple, S. & Quirke, N., Rapid imbibition of fluids in carbon nanotubes. *Phys. Rev. Lett.*, **90**, p. 214501, 2003.

[45] Humphrey, W., Dalke, A. & Schulten, K., VMD: Visual molecular dynamics. *J. Mol. Graph.*, **14**, p. 33, 1996.

[46] Martyna, G.J., Klein, M.L. & Tuckerman, M., Nosé–Hoover chains: The canonical ensemble via continuous dynamics. *J. Chem. Phys.*, **97**, p. 2635, 1992.

[47] Guillaud, E., Merabia, S., de Ligny, D. & Joly, L., Decoupling of viscosity and relaxation processes in supercooled water: A molecular dynamics study with the TIP4P/2005f model. *Phys. Chem. Chem. Phys.*, **19**, p. 2124, 2017.

[48] Chen, T., Smit, B. & Bell, A.T., Are pressure fluctuation-based equilibrium methods really worse than nonequilibrium methods for calculating viscosities?, *J. Chem. Phys.*, **131**, p. 246101, 2009.

[49] Nevins, D. & Spera, F.J., Accurate computation of shear viscosity from equilibrium molecular dynamics simulations. *Mol. Simul.*, **33**, p. 1261, 2007.

[50] Han, Y., Alsayed, A.M., Nobili, M., Zhang, J., Lubensky, T.C. & Yodh, A.G., Brownian motion of an ellipsoid. *Science*, **314**, p. 626, 2006.

[51] Happel, J. & Brenner, H., *Low Reynolds Number Hydrodynamics: With Special Applications to Particulate Media*, Kluwer Academic, 1983.

[52] Ferguson, A.L., Panagiotopoulos, A.Z., Debenedetti, P.G. & Kevrekidis, I.G., Systematic determination of order parameters for chain dynamics using diffusion maps. *Proc. Natl. Acad. Sci.*, **107**, p. 13597, 2010.

[53] Liu, Y., Wang, Q., Wu, T. & Zhang, L., Fluid structure and transport properties of water inside carbon nanotubes. *J. Chem. Phys.*, **123**, p. 234701, 2005.

[54] Joseph, S. & Aluru, N.R., Why are carbon nanotubes fast transporters of water?, *Nano Lett.*, **8**, p. 452, 2008.

[55] Barati Farimani, A. & Aluru, N.R., Spatial diffusion of water in carbon nanotubes: From fickian to ballistic motion. *J. Phys. Chem. B*, **115**, p. 12145, 2011.

[56] Köhler, M.H. & da Silva, L.B., Size effects and the role of density on the viscosity of water confined in carbon nanotubes. *Chem. Phys. Lett.*, **645**, p. 38, 2016.

[57] Zaragoza, A., Gonzalez, M.A., Joly, L., López-Montero, I., Canales, M.A., Benavides, A.L. & Valeriani, C., Molecular dynamics study of nanoconfined TIP4P/2005 water: How confinement and temperature affect diffusion and viscosity. *Phys. Chem. Chem. Phys.*, **21**, p. 13653, 2019.

[58] Mattia, D. & Gogotsi, Y., Review: Static and dynamic behavior of liquids inside carbon nanotubes. *Microfluid. Nanofluidics*, **5**, p. 289, 2008.

IMPLEMENTATION AND TESTING OF A NEW OPENFOAM SOLVER FOR PRESSURE-DRIVEN LIQUID FLOWS ON THE NANOSCALE

ALEXANDROS STAMATIOU, S. KOKOU DADZIE, & ALWIN M. TOMY
School of Engineering and Physical Sciences, Heriot-Watt University, Scotland, UK

ABSTRACT

Over the past two decades, several researchers have presented experimental data from pressure-driven water flow through carbon nanotubes quoting mass flow rates which are four to five orders of magnitude higher than those predicted by the Navier–Stokes equations with no-slip condition. The current work examines the development of an OpenFOAM solver for creeping flows that better accounts for some micro- and nano-scale diffusion processes. It is based on the observation that a change of velocity variable within the classical Navier–Stokes equations leads to a form of flow model with additional diffusive terms which become apparent at the micro- and nano-scale. Numerical simulations from the new solver compare well with associated analytical solutions that match the experimental flow enhancement observed in cylindrical tubes. This lays the foundations for further investigations of liquid flows in more complex nano-sized geometries, such as those obtained from pore-scale imaging.
Keywords: micro- and nanofluidics, continuum models, mass/volume diffusion, Navier–Stokes equations.

1 INTRODUCTION

Water transport in Carbon nanotubes (CNTs) has been a subject of intense research over the past two decades, predominantly because of its potential applications in technologies such as molecular level drug delivery and nanofiltration. The first major experimental studies of liquid flow through CNT membranes were carried out by Majumder et al. [1] and Holt et al. [2], on 7 nm diameter and 1.3–2 nm diameter CNTs respectively. Both investigations suggested extremely large water flow rate enhancements when compared to predictions based on the no-slip Haagen-Poiseuille flow law. In a repeat of the experiment, Majumder et al. [3] were later able to confirm their original findings of enhancement factors on the order of 10^3–10^4. In contrast to this, Qin et al. [4] reported flow enhancements of order 10^2–10^3 in 0.81–1.59 nm diameter CNTs, while the results by Du et al. [5] pointed to an enhancement of up to order 10^5 in 10 nm diameter channels. With the aim of expanding the data-set to include also wider tube diameters, Whitby et al. [6] investigated liquid flow through CNTs of 44 nm diameter, calculating only a 20–37 fold enhancement over no-slip Hagen-Poiseuille flow. For 200–300 nm diameter tubes, Sinha et al. [7] found no significant deviation, suggesting that flow enhancement effects diminish with increasing diameter.

A wide variety of flow enhancement data can also be found in the literature on molecular dynamics simulations of water flow through CNTs. For instance, Joseph and Aluru [8] measured an enhancement factor of 2052 in 2.17 nm diameter channels, while Thomas and McGaughey [9] found enhancements of 144–176 in 1.66–4.99 nm channels. Walther et al. [10] even report enhancements as low as 32, 25, 22 in simulations of 2.71, 4.07, 5.42 nm diameter tubes. Even if the spread in these data is large, the molecular dynamics simulations are consistent in the sense that none of them confirms the extremely high enhancement values found in some of the experimental studies. In fact, by employing the Young–Laplace equation, Walther et al. [10] study the water entry and filling stages of a CNT and derive a maximum attainable flow enhancement factor of 253. A similar upper bound is also argued

WIT Transactions on Engineering Sciences, Vol 132, © 2021 WIT Press
www.witpress.com, ISSN 1743-3533 (on-line)
doi:10.2495/MPF210071

for by Sisan and Lichter [11] using continuum methods. They show that frictional entrance and exit losses should not be neglected even for channels of small aspect ratio, thus limiting the flow rate.

It is common practice to quantify the flow enhancement using the slip velocity, which is a correction of the no-slip condition by the introduction of a constant velocity at the wall, leading to a slip-modified Haagen–Poiseuille mass flow rate. In experiments, the slip velocity is then found retrospectively by substituting the (enhanced) flow rate into this equation. A great deal of research has gone into effectively predicting the slip velocity a priori for a given fluid-solid combination. Using molecular dynamics simulations, this is often done by analysis of the parabolic velocity profile (see e.g. [9], [12]) or by use of the Navier friction coefficient (see [13], [14]). A different approach was suggested by Myers [15]. Leaving intact the no-slip condition, he modeled the flow enhancement by incorporating a region of reduced viscosity near the wall.

In Stamatiou et al. [16], a novel modelling approach was introduced in which the flow enhancement is caused by a diffusion mechanism that only becomes apparent at the nanoscale. A unifying recasting methodology was proposed by which a new class of continuum models termed Recast Navier–Stokes equations (RNS), can be directly derived from the Navier–Stokes equations [17], [18]. The idea is based on transforming the velocity vector field within the classical equations in a way that depends on the driving mechanism of the flow (for liquid flow in CNTs, this is the pressure gradient). The mass flow rate derived from this model as compared with experimental data showed reasonable agreement [16].

The objective of the current work is to present and validate a numerical implementation of the Recast Navier–Stokes equations in a new OpenFOAM solver (*rnsLiquidFoam*). For creeping flows in cylindrical tubes of small aspect ratio, a perturbation analysis yielding analytical expressions for the pressure and velocity fields are obtained [16]. These solutions are reviewed and used to validate the solver. The new solver can then be applied to simulate flows in complex geometries such as those obtained from pore-scale imaging techniques.

In Sections 2 and 3, the proposed equations for pressure-driven liquid flows and their numerical implementations are presented. In Sections 4 and 5, the perturbation solutions are reviewed and the numerical solutions are compared against them.

2 RECAST NAVIER–STOKES EQUATIONS

For a fluid of constant mass density (ρ), the Navier–Stokes mass and momentum conservation laws may be written as follows:

$$\nabla \cdot U_m = 0, \tag{1}$$

$$\frac{\partial}{\partial t}\left(\rho U_m\right) + \nabla \cdot \left(\rho U_m \otimes U_m\right) + \nabla \cdot \left[p\mathbf{I} + \Pi^{(NS)}\right] = 0. \tag{2}$$

Here, $\Pi^{(NS)}$ is the Newtonian stress tensor which is given in terms of the fluid's dynamic viscosity (μ) as:

$$\Pi^{(NS)} = -2\mu\left[\frac{1}{2}\left(\nabla U_m + (\nabla U_m)^T\right) - \frac{1}{3}\mathbf{I}\left(\nabla \cdot U_m\right)\right] = -2\mu\overline{\nabla U_m}^{\circ}. \tag{3}$$

The unknown velocity field $U_m = U_m(x, t)$ is that of the conventional mean *mass velocity*. If an externally applied pressure gradient is the principal driving mechanism of the flow, the new theory presented in Reddy et al. [17] and Dadzie and Reddy [18] assumes that the classical mass velocity can be written in terms of a new *pressure diffusion velocity* U_p as:

$$U_m = U_p - \kappa_p \nabla \ln p = U_p - \kappa_p \frac{\nabla p}{p}. \tag{4}$$

The second term on the right hand side represents a mass diffusion mechanism driven by the pressure gradient. This distinction between U_m and U_p is analogous to the idea of a volume velocity in gas [19]–[21]. The molecular *pressure diffusivity coefficient* κ_p is assumed to be of the form:

$$\kappa_p = \alpha^* \frac{\mu}{\rho} = \alpha^* \nu, \tag{5}$$

where α^* is a dimensionless parameter and ν the fluid's kinematic viscosity. Substituting eqn (4) into the Navier–Stokes eqns (1)–(2) and re-arranging terms leads to the recast Navier–Stokes (RNS) equations for mass and momentum:

$$\nabla \cdot U_p = \nabla \cdot (\kappa_p \nabla \ln p), \tag{6}$$

$$\frac{\partial}{\partial t} (\rho U_p - \kappa_p \rho \nabla \ln p) + \nabla \cdot (\rho U_p \otimes U_p) = \nabla \cdot \mathbf{T}^{(RNS)}, \tag{7}$$

with the tensor $\mathbf{T}^{(RNS)}$ on the right-hand side of this equation given by,

$$\mathbf{T}^{(RNS)} = \left(-p - \frac{2}{3} \frac{\mu \kappa_p}{p^2} |\nabla p|^2 + \frac{2}{3} \frac{\mu \kappa_p}{p} \nabla^2 p \right) \mathbf{I} + \frac{\kappa_p}{p^2} (2\mu - \rho \kappa_p) \nabla p \otimes \nabla p + 2\mu D (U_p)$$
$$- \frac{2}{3} \mu (\nabla \cdot U_p) \mathbf{I} - 2 \frac{\mu \kappa_p}{p} \nabla (\nabla p) + \frac{\rho \kappa_p}{p} U_p \otimes \nabla p + \frac{\rho \kappa_p}{p} \nabla p \otimes U_p. \tag{8}$$

Here, $D (U_p)$ denotes the symmetric part of the velocity gradient. A parallel can be drawn between the structure of tensor $\mathbf{T}^{(RNS)}$ and Korteweg's stress tensor $\mathbf{T}$ [22]. Korteweg augmented the Newtonian stress tensor with the dyadic product $\nabla \rho \otimes \nabla \rho$ to represent forces experienced by fluids during phase transitions. His complete tensor may be written as in [23]:

$$\mathbf{T} = \left(-p + \alpha_0 |\nabla \rho|^2 + \alpha_1 \nabla^2 \rho \right) \mathbf{I} + \beta \nabla \rho \otimes \nabla \rho + 2\mu D (\mathbf{v}) - \lambda (\nabla \cdot \mathbf{v}) \mathbf{I}, \tag{9}$$

where the material coefficients $\alpha_0, \alpha_1, \beta, \mu, \lambda$ may depend on ρ as well. On comparing eqn (9) with eqn (8), it is seen that all terms involved in the structure of the Korteweg stress tensor are found in the recast Navier–Stokes tensor, but written with p rather than ρ.

3 THE NEW OPENFOAM IMPLEMENTATION

The standard installation of OpenFOAM [24] comes with at least five incompressible solvers most of which are based on the PISO algorithm. In the incompressible Navier–Stokes equations, mass balance appears as a kinematic constraint on the velocity field (eqn (1)). There is no independent equation for the pressure, which presents a problem for the numerical computation of the solution. Solvers of the PISO family address this issue by constructing a Poisson equation for the pressure to enforce mass conservation, using one or more correction loops at each time-step [25]. This makes the solution of pressure expensive.

In the Recast Navier–Stokes setting, the pressure appears explicitly in the mass balance equation (eqn (6)). Instead of modifying one of the existing incompressible solvers, a simple sequential algorithm is proposed. This is shown in the source code of the new solver *rnsLiquidFlow* in Listing 1 below. The following steps may be identified here:

1. Define and compute the new diffusive term Jp and the surface flux of Jp (lines 1–3).
2. Define the part of the momentum equation involving the velocity Up (lines 6–12).

3. Solve the above momentum equation for velocity field using the pressure field from the previous time step (lines 14–19).
4. Define equation for pressure field (lines 21–24).
5. Solve for the pressure field using the previously calculated solution for velocity field (line 26).
6. Repeat the steps 1–5 for next time step.

```
1    lnp = log(p*dimensionedScalar("one", dimDensity/dimPressure, 1));
2    Jp = -alpha*nu*fvc::grad(lnp);
3    phiJp = fvc::flux(Jp);
4
5
6    fvVectorMatrix UEqn
7      (
8          fvm::ddt(Up)
9        + fvm::div(phi, Up)
10       - fvm::laplacian(nu, Up)
11       + fvm::div(phiJp, Up)
12      );
13
14   solve(UEqn == - fvc::grad(p)
15                 - fvc::ddt(Jp)
16                 + fvc::div(phi, Jp)
17                 + fvc::div(phiJp, Jp)
18                 + fvc::laplacian(nu, Jp)
19        );
20
21   fvScalarMatrix pEqn
22      (
23          fvm::laplacian(1.0/p, p)
24      );
25
26   solve(pEqn == fvc::div(Up)/(alpha*nu));
27   \
```

Listing 1: Excerpt of source code in rnsLiquidFoam.C

4 CREEPING FLOW IN CYLINDRICAL NANO-CHANNELS

4.1 Perturbation solutions

The tube-like channels embedded in a CNT membrane are characterised by a very small diameter D compared to their length L. It is therefore reasonable to seek a perturbation solution of eqns (6)–(7) in the channel aspect ratio $\varepsilon = D/L$, employing a cylindrical polar coordinate system (r, θ, z) in which the components of the pressure velocity are denoted U_{p_r}, U_{p_θ} and U_{p_z}. Fig. 1 shows a schematic of the fluid flow, which is assumed to be axisymmetric and with $U_{p_\theta} = 0$. The flow is driven by a pressure field p, such that $p(r,0) = P_{in}$ and $p(r, L) = P_{out}$ for all $0 \leq r \leq R$, where $P_{in} > P_{out}$ are the constant inlet and outlet pressures. On the channel wall, the tangential component of the velocity is assumed to satisfy the no-slip condition, i.e. $U_{p_z}(R, z) = 0$, so that any mass flow tangent to the wall must be purely diffusive. The radial component $U_{p_z}(R, z)$ on the other hand is related to the diffusive mechanism in such a way that no mass passes through the wall:

$$U_{m_r}(R, z) = U_{p_r}(R, z) - \kappa_p \frac{\partial \ln p}{\partial r}\bigg|_{(R,z)} = 0. \tag{10}$$

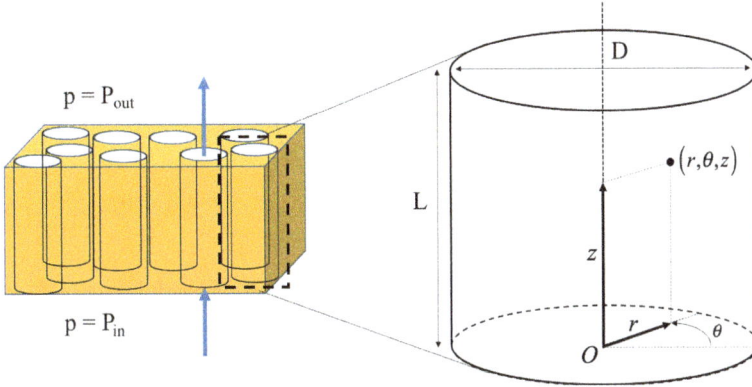

Figure 1: Schematic of CNT membrane and cylindrical geometry.

Eqn (10) suggests that some arbitrariness is left unless the gradient of the pressure at the wall is specified and this will be commented on in the solver validation section below.

In Table 1, dimensionless variables are defined in which the suggested perturbation solution is written as:

$$\tilde{p}(\tilde{r}, \tilde{z}) = \tilde{p}_0(\tilde{r}, \tilde{z}) + \varepsilon \tilde{p}_1(\tilde{r}, \tilde{z}) + \varepsilon^2 \tilde{p}_2(\tilde{r}, \tilde{z}) + \cdots, \tag{11}$$

$$\tilde{U}_{p_r}(\tilde{r}, \tilde{z}) = \tilde{U}_{p_r,0}(\tilde{r}, \tilde{z}) + \varepsilon \tilde{U}_{p_r,1}(\tilde{r}, \tilde{z}) + \varepsilon^2 \tilde{U}_{p_r,2}(\tilde{r}, \tilde{z}) + \cdots, \tag{12}$$

$$\tilde{U}_{p_z}(\tilde{r}, \tilde{z}) = \tilde{U}_{p_z,0}(\tilde{r}, \tilde{z}) + \varepsilon \tilde{U}_{p_z,1}(\tilde{r}, \tilde{z}) + \varepsilon^2 \tilde{U}_{p_z,2}(\tilde{r}, \tilde{z}) + \cdots. \tag{13}$$

For the dimensional analysis of different flow regimes the characteristic velocity magnitude $|U|_c$ is related to the characteristic pressure P_c via:

$$|U|_c = \frac{P_c D^2}{\mu L} = \frac{P_c \, \varepsilon H}{\mu}. \tag{14}$$

The Reynolds' number Re_p is calculated with respect to the pressure velocity U_p and can now be expressed as:

$$Re_p = \frac{|U|_c \, \rho D}{\mu} = \frac{P_c \, \rho \, \varepsilon D^2}{\mu^2}. \tag{15}$$

A Nusselt number Nu may be defined to compare the diffusive and advective transport mechanisms. With κ_p defined by eqn (5), this is directly related to the Reynolds' number:

$$Nu = \frac{\kappa_p}{UH} = \frac{\alpha^*}{Re_p}. \tag{16}$$

Table 1: Definition of dimensionless variables.

Dimensional variable	z	r	$U_{p,z}$	$U_{p,r}$	p
Scaling factor	L	D	$|U|_c$	$|U|_c$	P_c
Dimensionless variable	$\tilde{z} = \dfrac{z}{L}$	$\tilde{r} = \dfrac{r}{D}$	$\tilde{U}_{p,z} = \dfrac{U_{p,z}}{|U|_c}$	$\tilde{U}_{p,r} = \dfrac{U_{p,r}}{|U|_c}$	$\tilde{p} = \dfrac{p}{P_c}$

4.2 Flow regime with $\mathrm{Re_p} = O(\varepsilon^2)$ and $\alpha^* = O(1)$

Table 2 in Section 5 lists a set of parameters typical for flow experiments carried out in nano-diameter channels of small aspect ratio. Based on these values, the dimensionless numbers $\varepsilon = 1 \times 10^{-3}$ and $\mathrm{Re_p} = 1 \times 10^{-5}$ are found. The low Reynolds' number is typical of creeping flows so that inertial terms can be neglected in the momentum equation. A rigorous perturbation analysis on the non-dimensionalised equations with $\mathrm{Re_p} = O(\varepsilon^2)$ and $\alpha^* = O(1)$ reveals that the first two pressure terms in eqn (11) are independent of the radial coordinate r (see [16] for details). In other words, $\tilde{p}_0 = \tilde{p}_0(z)$ and $\tilde{p}_1 = \tilde{p}_1(z)$. Using this and equating the $O(1)$ terms in the mass balance equation results in:

$$\tilde{p}_0^2 \frac{1}{\tilde{r}} \frac{\partial \left(\tilde{r} \, \tilde{U}_{\mathrm{p_r},0} \right)}{\partial \tilde{r}} = \frac{\alpha^*}{\mathrm{Re_p}} \left[\varepsilon^2 \tilde{p}_0 \frac{d^2 \tilde{p}_0}{d\tilde{z}^2} - \varepsilon^2 \left(\frac{d\tilde{p}_0}{d\tilde{z}} \right)^2 + \tilde{p}_0 \frac{1}{\tilde{r}} \frac{\partial}{\partial \tilde{r}} \left(\tilde{r} \frac{\partial \tilde{p}_2}{\partial \tilde{r}} \right) \right]. \quad (17)$$

It should be noted here that, within the perturbation analysis, the $O(1)$ term of the radial component of the mass velocity (see eqn (10)) is:

$$\tilde{p}_0 \tilde{U}_{\mathrm{m_r},0} = \tilde{p}_0 \tilde{U}_{\mathrm{p_r},0} - \frac{\alpha^*}{\mathrm{Re_p}} \frac{\partial \tilde{p}_2}{\partial \tilde{r}}. \quad (18)$$

The pressure term $\tilde{p}_0(\tilde{z})$ can be found directly by integrating eqn (17) and using the boundary condition $\tilde{U}_{\mathrm{m_r},0} = 0$ at the wall. This leads to the following ordinary differential equation:

$$\tilde{p}_0 \frac{d^2 \tilde{p}_0}{d\tilde{z}^2} - \left(\frac{d\tilde{p}_0}{d\tilde{z}} \right)^2 = 0 \quad (19)$$

Introducing the ratio $\mathcal{P} = \mathrm{P_{in}}/\mathrm{P_{out}}$, the dimensional form of the solution is:

$$p_0(z) = \mathrm{P_{in}} \exp \left(-\frac{\ln(\mathcal{P})z}{L} \right) \quad \Longleftrightarrow \quad \frac{z}{L} \ln \mathcal{P} + \ln \frac{p_0(z)}{\mathrm{P_{in}}} = 0. \quad (20)$$

Furthermore, equating the $O(1)$ terms in the z-momentum equation yields:

$$\frac{\partial^2 \tilde{U}_{\mathrm{p_z},0}}{\partial \tilde{r}^2} + \frac{1}{\tilde{r}} \frac{\partial \tilde{U}_{\mathrm{p_z},0}}{\partial \tilde{r}} = \frac{d\tilde{p}_0}{d\tilde{z}}. \quad (21)$$

With application of the no-slip condition on $\tilde{U}_{\mathrm{p_z},0}$, eqn (21) results in a parabolic velocity profile, the dimensional form of which is:

$$\mathrm{U}_{\mathrm{p_z},0}(r, z) = \frac{1}{4\mu} \left(r^2 - R^2 \right) \frac{dp_0}{dz}. \quad (22)$$

It should be emphasized that, in the high Nusselt number flow regime considered here, the stream-wise velocity term $\mathrm{U}_{\mathrm{p_z},0}$ does not feature in eqn (17). The diffusive transport dominates and determines the pressure distribution with help of the no-penetration condition at the wall.

4.3 Flow regime with $\mathrm{Re_p} = O(\varepsilon^2)$ and $\alpha^* = O(\varepsilon)$

In Stamatiou et al. [16], the case with $\mathrm{Re_p} = O(\varepsilon)$ and $\alpha^* = O(1)$ was investigated. However, closer inspection reveals that the same terms dominate the equations for the flow regime in which $\mathrm{Re_p} = O(\varepsilon^2)$ and $\alpha^* = O(\varepsilon)$. This is because the Nusselt number $\mathrm{Nu} = \alpha^*/\mathrm{Re_p}$ is the same for both these regimes and the terms proportional to the Reynolds number are negligible. In comparison to the case examined in Section 4.2, there is one significant change: while the z-momentum equation is again given by eqn (21), the mass balance equation now exhibits an extra term coupling the two equations together:

$$\tilde{p}_0^2 \frac{\partial \tilde{U}_{\mathrm{Pz},0}}{\partial \tilde{z}} + \tilde{p}_0^2 \frac{1}{\tilde{r}} \frac{\partial \left(\tilde{r}\, \tilde{U}_{\mathrm{Pr},0} \right)}{\partial \tilde{r}} = \frac{\alpha^*}{\mathrm{Re_p}} \left[\varepsilon^2 \tilde{p}_0 \frac{d^2 p_0}{d\tilde{z}^2} - \varepsilon^2 \left(\frac{d\tilde{p}_0}{d\tilde{z}} \right)^2 + \tilde{p}_0 \frac{1}{\tilde{r}} \frac{\partial}{\partial \tilde{r}} \left(\tilde{r}\, \frac{\partial \tilde{p}_2}{\partial \tilde{r}} \right) \right].$$
(23)

The pressure term $\tilde{p}_0(\tilde{z})$ is now found by substituting the dimensionless form of eqn (22) in eqn (23), integrating and using $\tilde{U}_{\mathrm{m_r},0} = 0$ at the wall. The resulting ordinary differential equation for $\tilde{p}_0(\tilde{z})$ now becomes:

$$\frac{d^2 \tilde{p}_0}{d\tilde{z}^2} + \frac{32\, \alpha^* \varepsilon}{\mathrm{Re_p}} \frac{d^2 \ln \tilde{p}_0}{d\tilde{z}^2} = 0.$$
(24)

This equation only admits an implicit solution, the dimensional form of which can be written as follows:

$$p_0(z) + \frac{8\mu\kappa_\mathrm{p}}{R^2} \left[\frac{z}{L} \ln \mathcal{P} + \ln \frac{p_0(z)}{\mathrm{P_{in}}} \right] = \mathrm{P_{in}} + \frac{\Delta P}{L} z,$$
(25)

where $\mathcal{P} = \mathrm{P_{in}}/\mathrm{P_{out}}$ as before and $\Delta P = \mathrm{P_{out}} - \mathrm{P_{in}}$ is the pressure drop. Written in this form, it can be seen that the pressure distribution for the present flow regime is a correction of eqn (20) by a linear pressure drop. Eqn (20) is approximated for large values of $\mu\kappa_\mathrm{p}/R^2$, while the Haagen-Poiseuille flow law is recovered for small values of $\mu\kappa_\mathrm{p}/R^2$. This is also seen when considering the mass flow rate expression:

$$\dot{\mathrm{M}}_{\mathrm{RNS}} = \frac{\pi\rho R^4}{8\mu L} \left[\Delta P + \frac{8\mu\kappa_\mathrm{p}}{R^2} \ln\left(\mathcal{P} \right) \right] = \dot{\mathrm{M}}_{\mathrm{HP}} \left(1 + E_{\mathrm{RNS}} \right),$$
(26)

where $\dot{\mathrm{M}}_{\mathrm{HP}}$ is the no-slip Haagen-Poiseuille mass flow rate, i.e.

$$\dot{\mathrm{M}}_{\mathrm{HP}} = \frac{\pi\rho R^4 \Delta P}{8\mu L},$$
(27)

and E_{RNS} is the flow enhancement factor defined as,

$$E_{\mathrm{RNS}} = \frac{\dot{\mathrm{M}}_{\mathrm{RNS}} - \dot{\mathrm{M}}_{\mathrm{HP}}}{\dot{\mathrm{M}}_{\mathrm{HP}}} = \frac{8\mu\kappa_\mathrm{p} \ln\left(\mathcal{P} \right)}{R^2 \Delta P}.$$
(28)

5 SOLVER VALIDATION

5.1 Computational mesh and solver settings

In this section it is verified that the *rnsLiquidFoam* solver agrees with the previously found analytical expressions for flows in cylindrical tubes. The axial symmetry of the problem permits the solution of the equations on a small wedge instead of the entire cylinder. Fig. 2 shows the computational mesh for a wedge angle of $5°$ with 100 uniformly spaced cells in the z-direction and 20 cells in the y-direction with an expansion ratio of 0.1, so that the mesh becomes more refined in the near-wall region. The small angle allows the wall to be approximated by a flat boundary patch with outward normal $\hat{n}$ in the y-direction. On this boundary, the conditions $U_p = (0, 0, 0)$ and $\nabla p \cdot \hat{n} = 0$ (zeroGradient) are adopted, which is stricter than the boundary condition used to derive the first order perturbation solutions in Section 4 (see eqn (10)). The angled top and bottom planes of the mesh are equipped with the wedge condition for rotationally symmetric cases. Table 2 summarises the other physical parameters used for the simulation of water flow in a CNT of 10 nm diameter and 10 μm length driven by a pressure difference of 1 bar.

5.2 Comparison with analytical solutions

In Section 4.3 the influence of the diffusivity parameter α^* on the pressure distribution was highlighted. This is illustrated in Fig. 3, where eqn (25) is plotted for $\alpha^* = 1$ and $\alpha^* = 0.001$, representing extremal values for a large range of possible diffusivities κ_p. The same curves are also computed with *rnsLiquidFoam*, sampling along the z-axis. Excellent agreement can be seen in both cases. On the one hand, the case with $\alpha^* = 1$ corresponds to a flow process dominated by the diffusive transport mechanism, the convective transport being negligible. On the other hand, the case with $\alpha^* = 0.001$ leads to both mechanisms contributing equally. As α decreases, the pressure distribution is seen to approach the linear pressure drop of classical Poiseuille flow. For illustrative purposes, the value $\alpha^* = 0.005$ is used in all subsequent plots, which represents a case where diffusive transport is of the same

Table 2: Parameters for simulation of water flow in CNTs.

Parameter	L (m)	D (m)	μ (kg m^{-1}s^{-1})	ρ (kg m^{-3})	P_{in} (Pa)	P_{out} (Pa)
Value	1×10^{-5}	1×10^{-8}	1×10^{-3}	1×10^3	2×10^5	1×10^5

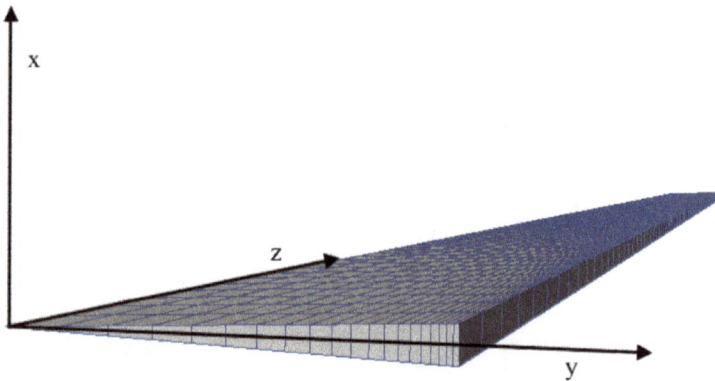

Figure 2: Computational mesh for a wedge of $5°$.

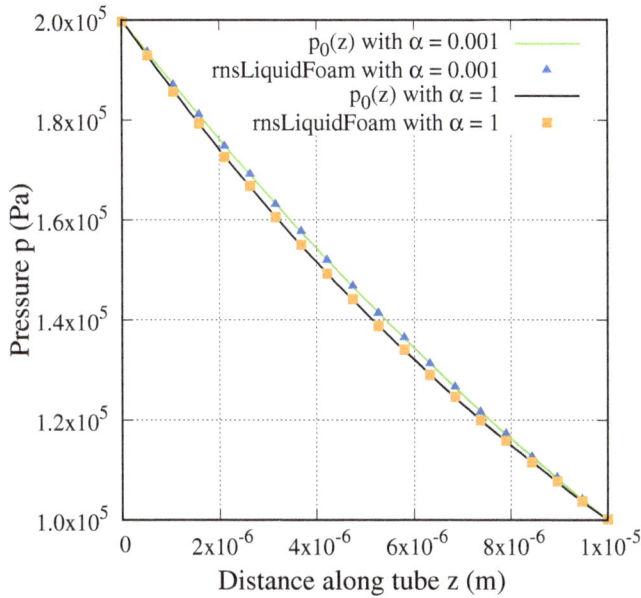

Figure 3: Pressure distribution along the tube.

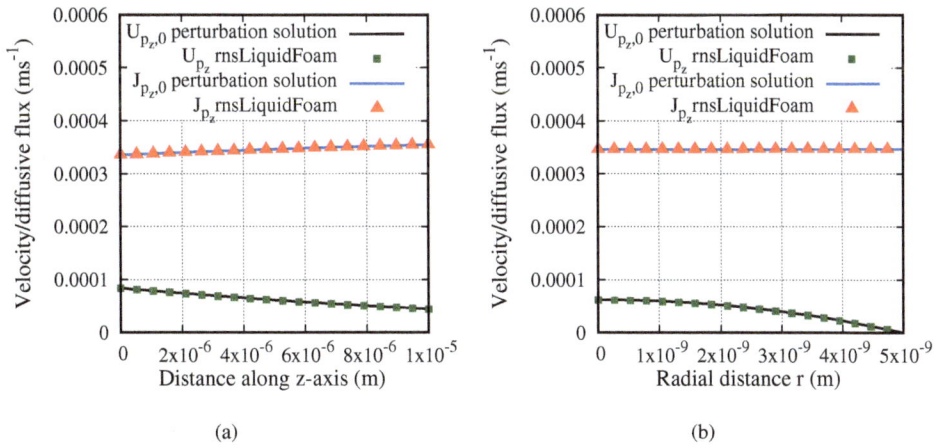

Figure 4: Stream-wise velocity and diffusion components on two axes.

order but clearly higher than convective transport. Fig. 4(a) shows the variation of stream-wise components U_{p_z} and J_{p_z} on the z-axis.

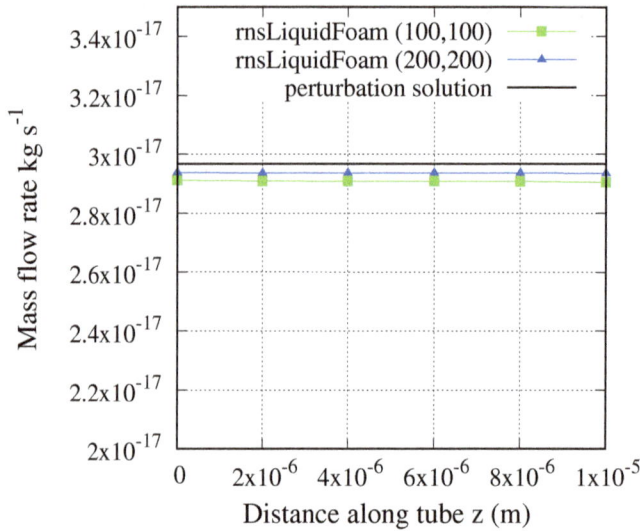

Figure 5: Mass flow rate calculation on two grids.

Both are non-constant, owing to the choice of α^* (not too 'small' or 'large'). Fig. 4(b) shows the same components plotted against the radial distance from the z-axis at the centre of the channel ($z = 5 \times 10^{-9}$m). The numerical solution of U_{p_z} agrees well with the parabolic profile $U_{p_{z,0}}$ (eqn (22)) and J_{p_z} is constant. The mass velocity profile U_{m_z} is therefore also parabolic and indistinguishable from that obtained using a velocity slip condition in the classical Navier–Stokes equations.

Fig. 5 shows the calculation of the mass flow rate through six representative cross sections of the entire tube. This was initially done using a 100×100 grid (green squares) and then repeated on a 200×200 (blue triangles) refined grid. The mass flow rate is constant along the channel in both numerical calculations and the solution appears to converge to eqn (26). A deviation with respect to this analytical expression must persist also because the circular tube wall has been approximated by a polygonal wall.

Fig. 6 shows the flow enhancement for a selection of experimental studies and molecular dynamics simulations of water flow in CNTs, which differ both in channel dimensions and in imposed pressure differences. Eqn (28) is plotted (orange triangles connected by straight line segments) using the dimensions and pressures quoted in each individual investigation. A geometry-dependent form of α^* has been employed here, given by $\alpha^* = 0.003 \times L/D$ [16]. The corresponding simulation results from the new solver, *rnsLiquidFoam*, are shown by the blue open circles connected with dashed straight line segments.

It should be noted that molecular dynamics simulations have revealed that the radial pressure distribution in nano-sized tubes is not constant, dropping off sharply near the boundaries [10]. Similarly, the viscosity is slightly reduced in this zone, which has been incorporated in previous modelling attempts [15]. The kinematic viscosity $\nu = \mu/\rho$ enters as

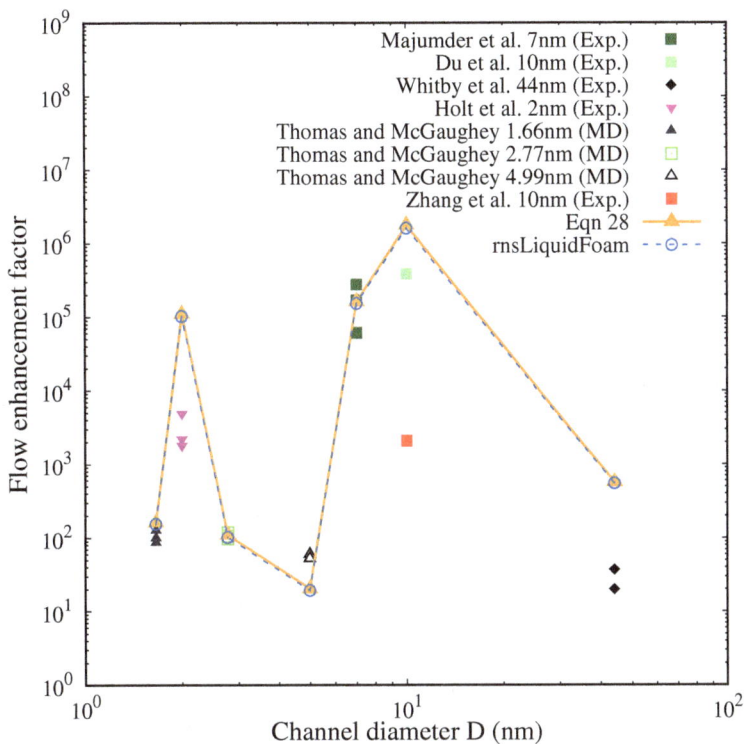

Figure 6: Comparison between numerically simulated, experimental and analytical flow rate enhancement.

a constant in the new model and considering its radial variation could lead to an improved agreement with the data. Also the inclusion of end-effects should be examined, as suggested in Sisan and Lichter [11].

6 CONCLUSIONS

This paper has explored the numerical implementation in OpenFOAM of a Recast Navier–Stokes equations for pressure-driven liquid flows. The explicit appearance of the pressure in the mass balance equation suggests a simplified numerical solution process compared to that of the typical incompressible Navier–Stokes equations where mass balance must be enforced by a Poisson equation for the pressure and several correction steps. For liquid flow in CNTs, perturbation solutions in the small channel aspect ratio were reviewed and used to test convergence of the new solver. Very good agreement was obtained for the pressure distribution and velocity profiles. The calculation of the mass flow rate showed a small deviation from the analytical expression, but appeared to converge to it upon grid-refinement. We also showed that the solver computes the correct mass flow rates for the different channel sizes and pressure differences used in experimental and molecular dynamics studies.

ACKNOWLEDGEMENTS

This research is supported by the UK's Engineering and Physical Sciences Research Council (EPSRC) under grant no. EP/R008027/1 and The Leverhulme Trust, UK, under grant Ref. RPG-2018-174.

REFERENCES

[1] Majumder, M., Chopra, N., Andrews, R. & Hinds B.J., Enhanced flow in carbon nanotubes. *Nature*, **438**, p. 44, 2005.

[2] Holt, J.K., Park, H.G., Wang, Y., Stadermann, M., Artyukhin, A.B., Grigoropoulos, C.P., Noy, A. & Bakajin, O., Fast mass transport through sub2-nanometer carbon nanotubes. *Science*, **312**, pp. 1034–1037, 2006.

[3] Majumder, M., Chopra, N., Andrews, R. & Hinds, B.J., Mass transport through carbon nanotube membranes in three different regimes: Ionic diffusion and gas and liquid flow. *Nano*, **5**, pp. 3867–3877, 2011.

[4] Qin, X., Yuan, Q., Zhao, Y., Xie, S. & Liu, Z., Measurement of the rate of water translocation through carbon nanotubes. *Nano Lett.*, **11**, pp. 2173–2177, 2011.

[5] Du, F., Qu, L., Xia, Z., Feng, L. & Dai, L., Membranes of vertically aligned superlong carbon nanotubes. *Langmuir*, **27**, pp. 8437–8443, 2011.

[6] Whitby, M., Cagnon, L., Thanou, M. & Quirke, N., Enhanced fluid flow through nanoscale carbon pipes. *Nano Lett.*, **8**, pp. 2632–2637, 2008.

[7] Sinha, S., Rossi, M. P., Mattia, D. & Godotsi, Y., Induction and measurement of minute flow rates through nanopipes. *Phys. Fluids*, **19**, 013603, 2007.

[8] Joseph, S. & Aluru, N.R., Why are carbon nanotubes fast transporters of water? *Nano Lett.*, **8**, pp. 452–458, 2008.

[9] Thomas, J.A. & McGaughey, A.J.H., Reassessing fast water transport through carbon nanotubes. *Nano Lett.*, **8**, pp. 2788–93, 2008.

[10] Walther, J.H., Ritos, K., Cruz-Chu, E.R., Megaridis, C.M. & Koumoutsakos, P., Barriers to superfast water transport in carbon nanotube membranes. *Nano Lett.*, **13**, pp. 1910–1914, 2013.

[11] Sisan, T.B. & Lichter, S., The end of nanochannels. *Microfluid Nanofluid*, **11**, pp. 787–791, 2011.

[12] Kannam, S.K., Todd, B.D., Hansen, J.S. & Daivis, P.J., How fast does water flow in carbon nanotubes? *J. Chem. Phys.*, **138**, 094701, 2013.

[13] Hansen, J.S., Todd, B.D. & Daivis, P.J., Prediction of fluid velocity slip at solid surfaces. *Phys. Rev. E*, **84**, 016313, 2011.

[14] Kannam, S.K., Todd, B.D., Hansen, J.S. & Daivis, P.J., Interfacial slip friction at a fluid-solid cylindrical boundary. *J. Chem. Phys.*, **136**, 244704, 2012.

[15] Myers, T.G., Why are slip lengths so large in carbon nanotubes? *Microfluidics and Nanofluidics*, **10**(5), pp. 1141–1145, 2010.

[16] Stamatiou, A., Dadzie, S.K. & Reddy, M.H.L., Investigating enhanced mass flow rates in pressure-driven liquid flows in nanotubes. *J. Phys. Commun.*, **3**, p. 125012, 2019.

[17] Reddy, M.H.L., Dadzie, S.K., Ocone, R., Borg, M.K. & Reese, J.M., Recasting Navier-Stokes equations. *J. Phys. Commun.*, **3**, p. 105009, 2019.

[18] Dadzie, S.K. & Reddy, M.H.L., Recasting Navier-Stokes equations: Shock wave structure description. *AIP Conference Proceedings*, **2293**, p. 050005, 2020.

[19] Dadzie, S.K., Reese, J.M. & McInnes, C.R., A continuum model of gas flows with localized density variations. *Physica A*, **387**, pp. 6079–6094, 2008.

[20] Brenner, H., Beyond Navier-Stokes. *Int. J. Eng. Sci.*, **54**, pp. 67–98, 2012.

[21] Dadzie, S. K. & Brenner, H., Predicting enhanced mass flow rates in gas microchannels using nonkinetic models. *Phys. Rev. E*, **86**, 2012.

[22] Korteweg, D.J., Sur la forme que prennent les equations du mouvements des fluides si l'on tient compte des forces capillaires causees par des variations de densite considerables mais connues et sur la theorie de la capillarite dans l'hypothese d'une variation continue de la densite. *Arch. Neerl. Sci. Ex. Nat. (ii) 6*, 1901.

[23] Heida, M. & Malek, J., On compressible Korteweg fluid-like materials. *Int. J. Eng. Sci., special Issue in Honor of K.R. Rajagopal.*, **48**(11), pp. 1313–1324, 2010.

[24] Weller, H.G., Tabor, G., Jasak, H. & Fureby, C., A tensorial approach to computational continuum mechanics using object-oriented techniques. *Computers in Physics*, **12**, p. 620, 1998.

[25] Ferziger, J.H. & Peric, M., *Computational Methods for Fluid Dynamics*. Springer-Verlag: Berlin, Heidelberg, New York, 2002.

CONSIDERATION OF THE MICROSTRUCTURE OF PARTICLE SUSPENSION TO ESTIMATE ITS INTRINSIC VISCOSITY

TOMOHIRO FUKUI, MISA KAWAGUCHI & KOJI MORINISHI
Department of Mechanical Engineering, Kyoto Institute of Technology, Japan

ABSTRACT
It is important to comprehend rheology of particle suspension to facilitate useful and effective applications in many fields. It is well known that relative viscosity for higher concentration becomes higher than that form Einstein's viscosity equation. One of the reasons is interaction between suspended particles and suspending fluids. We previously considered the influence of interactions on increase in relative viscosity by focusing on the suspended particles' rotational motions. For higher concentrated suspensions, the rotational motions were disturbed because particles were too jammed to move freely, especially in rotational direction. Since these rotational motions were strongly related to the total macroscopic fluid resistance, the relative viscosity depended on the microscopic gaps between the particles. In the meantime, relationships between microstructure, i.e., spatial arrangement of the particles, and rheology are still unclear. In this study, therefore, numerical simulations were conducted to consider relative and intrinsic viscosities in terms of microstructure of suspensions. The results showed that the concentration profile of the particle suspension was almost flat except for near the channel wall. Few particles flowed near the wall because repulsive force from the wall increased exponentially with approaching the wall. They flowed away from the channel wall. For the higher concentrated suspension, on the other hand, particles were too jammed to flow smoothly near the channel center. Then some particles were pushed out toward the channel wall against the repulsive force. In this case, they flowed near the channel wall as well. Owing to these effects, the concentration profile for the concentrated suspension depicted almost flat including near the wall. The microstructure for higher concentrated suspension also changed with changes in concentration profile. Then the relative and intrinsic viscosities were consequently increased with concentration. The intrinsic viscosity was significantly related to the microstructure of the suspension.
Keywords: rheology, non-Newtonian property, dilute suspension, microstructure, margination, two-way coupling simulation.

1 INTRODUCTION

It is important to comprehend rheology of particle suspension to facilitate useful and effective applications in many fields. One of the most convenient aspects by using particle suspension is its simple relationship between particle concentration and consequent apparent viscosity. According to Einstein's viscosity equation [1], the apparent effective viscosity η_{eff} can be simply related to the particle concentration ϕ with intrinsic viscosity [η]:

$$\eta_{\text{eff}} = \eta_0\{1 + [\eta]\phi\}, \tag{1}$$

where η_0 is the viscosity of the solvent. The intrinsic viscosity [η] depends on the shape of the suspended particles: for the case of spherical particles, [η] = 2.0 for two-dimensional [2] and [η] = 2.5 for three-dimensional [1]. In order to compare contributions of suspended particles to the apparent viscosity among suspensions with different solvent, relative viscosity η_{eff}/η_0 is sometimes preferably employed:

WIT Transactions on Engineering Sciences, Vol 132, © 2021 WIT Press
www.witpress.com, ISSN 1743-3533 (on-line)
doi:10.2495/MPF210081

$$\frac{\eta_{\text{eff}}}{\eta_0} = 1 + [\eta]\phi. \tag{2}$$

It is well known that relative viscosity for higher concentration becomes higher than that form Einstein's viscosity equation. One of the reasons is interaction between suspended particles and suspending fluids [3]–[6]. We previously considered the influence of interactions on increase in relative viscosity by focusing on the suspended particles' rotational motions [7], [8]. For higher concentrated suspensions, the rotational motions were disturbed because particles were too jammed to move freely, especially in rotational direction. Since these rotational motions were strongly related to the total macroscopic fluid resistance [7], the relative viscosity, i.e., pressure loss mainly due to viscous dissipation, depended on the microscopic gaps between the particles.

Recently, effects of non-Newtonian properties of the solvent have also received much attention. In a non-Newtonian solvent, since macroscopic velocity profiles strongly depend on the shear rate, suspended particles' behaviors are entirely different from those in a Newtonian solvent. For example, Hu et al. [9] and Christ et al. [10] showed the preferable radial equilibrium positions for the suspended particles in non-Newtonian fluids. Tanaka et al. [11] reported effects of the power-law fluidic properties on the suspension rheology. They successfully showed increase in relative and intrinsic viscosities of a suspension attributed by the non-Newtonian solvent. These findings are important especially in a field of bioengineering. It is reported that blood from patients suffering from cardiovascular disease included much more proteins within the plasma [12]. This may lead to higher viscosities of blood. They also showed higher death rate related to large amount of proteins in plasma. It is important to consider mechanisms of viscosity changes due to interactions between suspending fluid and suspended particles.

In the meantime, relationships between microstructure, i.e., spatial arrangement of the particles [13], and rheology are still unclear. Doyeux et al. [6] considered effects of particle's radial position on the total effective viscosity. They showed that when a particle approached the channel wall, the effective viscosity increased exponentially. Thus, the effective viscosity is not only a function of concentration ϕ but also strongly influenced by its microstructure. They also proposed an alternative estimation for effective viscosity considering its microstructure instead of Einstein's equation. Based on their proposal, Okamura et al. [14] recently validated total relative viscosity of a suspension by considering summation of each particle's contribution. They showed that relative viscosity could be estimated by its microstructure in a limited condition. Although their study was still preliminary, viscosity estimation by its microstructure would be a promising approach and more considerations should be necessary. It is also expected to reveal the mechanism of viscosity changes by considering microstructure of a suspension. Fukui et al. [15] showed microstructure changes due to inertial effects of the suspended particles and consequent viscosity decrease of suspension. These microstructure changes were considered to be one of the major factors in changing macroscopic rheology. On the other hand, since these inertial effects generally tended to cause thixotropic behavior, i.e., viscosity decrease, it is important to consider microstructure changes resulting in viscosity increase as discussed above. In this study, we focus on the microstructure in a pressure-driven suspension flow with different concentrations in order to consider mechanism of viscosity increase with increasing concentration. We also consider the relationship between microstructure and its relative and intrinsic viscosities.

2 METHODS

2.1 Computational models

We conducted two-dimensional pressure-driven suspension flow simulations by a two-way coupling scheme. Fig. 1 shows a schematic view of the simulation model used in this study. The cannel width was $2l = 400$ μm, and the axial length was set 4 times as long as its width. A periodic boundary condition considering pressure was applied axial direction to reduce computational costs. Suspended particles with a diameter $2r = 20$ μm were randomly distributed as an initial condition. Note that at least 20 μm gaps between the particles or particle-wall were allowed for the initial position to stabilize the computation. Number of particles was set 21 for $\phi = 1.02\%$, 42 for $\phi = 2.04\%$, and 84 for $\phi = 4.07\%$, respectively. The simulations were then conducted until physical time $t = 8$ s, which corresponds to non-dimensional time of 100. More or less, initial random positions of the particles affect particles flow patterns and consequent microstructure [15], these simulations were repetitively carried out 40 times for $\phi = 1.02\%$, and 20 times for $\phi = 2.04$ and 4.07%, respectively. The spatial resolution was set 1 μm for both directions, which has been validated by grid independence test [8]. Particle shape was described by virtual flux method [16] to satisfy the hydrodynamic boundary conditions on the particle surface in Cartesian coordinate system with regular intervals.

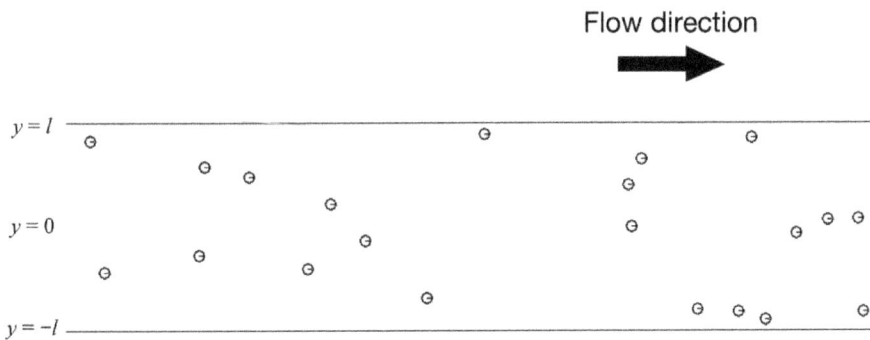

Figure 1:   Schematic view of a pressure-driven suspension flow model. The channel width and length were set 400 μm and 1,600 μm, respectively. Periodic boundary condition was applied in the x direction. Suspended particles with a diameter $2r = 20$ μm were randomly distributed as an initial condition. Number of particles was set 21 for $\phi = 1.02\%$, 42 for $\phi = 2.04\%$, and 84 for $\phi = 4.07\%$, respectively.

2.2 Governing equation for suspending fluid

The governing equation for suspending fluid was regularized lattice Boltzmann equation [17], [18], which is a modified form of the original lattice Boltzmann equation in order to stabilize the computation. Briefly, distribution function f_α in the regularized lattice Boltzmann equation is written by using up to second-order moments:

$$f_\alpha = \omega_\alpha \left(a_0 + b_i e_{\alpha i} + c_{ij} e_{\alpha i} e_{\alpha j}\right), \tag{3}$$

where ω is the weight factor, α is the direction of the discrete velocity vector e_α, and a_0, b_i, c_{ij} are the parameters that satisfy the following relationships:

$$\sum_\alpha f_\alpha = \rho, \tag{4}$$

$$\sum_\alpha e_{\alpha i} f_\alpha = \rho u_i, \tag{5}$$

$$\sum_\alpha e_{\alpha i} e_{\alpha j} f_\alpha = \frac{c^2}{3}\rho\delta_{ij} + \rho u_i u_j + \Pi_{ij}^{\text{neq}}, \tag{6}$$

where Π_{ij}^{neq} is the nonequilibrium part of the stress tensor. The distribution function f_α in eqn (3) is then,

$$f_\alpha = \omega_\alpha \rho \left[1 + \frac{3(e_{\alpha i}u_i)}{c^2} + \frac{9(e_{\alpha i}u_i)^2}{2c^4} - \frac{3(u_i u_i)}{2c^2}\right] + \frac{9\omega_\alpha}{2c^2}\left(\frac{e_{\alpha i}e_{\alpha j}}{c^2} - \frac{1}{3}\delta_{ij}\right)\Pi_{ij}^{\text{neq}}. \tag{7}$$

The first term is equivalent to Maxwell equilibrium distribution function f_α^{eq} with low Mach number approximation. When the distribution function f_α can be expanded by a power series of Kundsen number ε around the equilibrium distribution function f_α^{eq}, the distribution function f_α is written as

$$f_\alpha = f_\alpha^{\text{eq}} + f_\alpha^{\text{neq}} = f_\alpha^0 + f_\alpha^1 + f_\alpha^2 + \cdots, \tag{8}$$

where f_α^n corresponds to of the order of $O(\varepsilon^n)$, and f_α^0 is equal to f_α^{eq}. Therefore, the second term of eqn (7) can be substituted for f_α^1, and the time evolution equation for the regularized lattice Boltzmann equation is

$$f_\alpha(t + \Delta t, x + e_\alpha\Delta t) = f_\alpha^{\text{eq}}(t, x) + \left(1 - \frac{1}{\tau}\right)f_\alpha^1(t, x), \tag{9}$$

where τ is the relaxation time. When Navier–Stokes equations are derived from lattice Boltzmann equation through Chapman–Enskog expansion procedure, the relaxation time τ is defined as follows in the incompressible limit [19],

$$\tau = \frac{3v}{c\Delta x} + \frac{\Delta t}{2}, \tag{10}$$

where v is the kinematic viscosity. The relaxation time $\tau = 0.74$ for all our computations.

2.3 Governing equations for suspended particles

The suspended particles used in this study were assumed to be rigid, spherical, chemically stable, and non-Brownian. Their movements were simply described by Newton's second law of motion and equation of angular motion:

$$F_p = \rho\frac{d^2 x_p}{dt^2}, \tag{11}$$

$$T_p = I\frac{d^2 \theta_p}{dt^2}, \tag{12}$$

where F_p is the external hydrodynamic force vector acting on the particle, ρ is the density, x_p is the position vector, T_p is the torque, I is the moment of inertia, and θ_p is the angle of the particle. Note that both densities of suspended particles and suspending fluid were assumed to be equivalent for neutral buoyancy. The external force vector F_p and torque T_p acting on the particles were discretized by a third-order Adams–Bashforth method and solved numerically by a two-way coupling scheme [7]:

$$\dot{x}_p^{n+1} = \dot{x}_p^n + \Delta t \frac{23F_p^n - 16F_p^{n-1} + 5F_p^{n-2}}{12\rho}, \tag{13}$$

$$x_p^{n+1} = x_p^n + \Delta t \frac{5\dot{x}_p^{n+1} + 8\dot{x}_p^n - \dot{x}_p^{n-1}}{12}, \tag{14}$$

$$\dot{\theta}_p^{n+1} = \dot{\theta}_p^n + \Delta t \frac{23T_p^n - 16T_p^{n-1} + 5T_p^{n-2}}{12I}, \tag{15}$$

$$\theta_p^{n+1} = \theta_p^n + \Delta t \frac{5\dot{\theta}_p^{n+1} + 8\dot{\theta}_p^n - \dot{\theta}_p^{n-1}}{12}. \tag{16}$$

3 RESULTS AND DISCUSSION

The macroscopic axial velocity and microscopic particles behavior at $t = 8$ s are depicted in Fig. 2. The suspended particles are shown to flow randomly in a channel. Note that since sufficient number of grids was allocated in the computational domain, collisions between particles or particle and channel wall were not observed. At least 2 or 3 grids always remained between the particles during the simulations. Since inertial effects were negligible due to low Reynolds number condition, the particles did not migrate in the width direction, i.e., the particles flowed almost along the macroscopic streamline. The particles, however, did not flow very near the channel wall. The particles apparently flowed avoiding a certain peripheral layer near the channel wall in Fig. 2(a) and 2(b). This is because repulsive force from the wall increased exponentially with approaching the wall [6], [14]. Owing to these effects, the particles did not approach the wall and some layers without particles existed near the wall. However, for the case $\phi = 4.07\%$, particles flowed within these regions too as shown in Fig. 2(c). This is partly because when particles are getting jammed with increasing concentration, some particles are pushed away toward the peripheral layers near the wall. These phenomena are sometimes observed in microcirculation termed "margination". Deformable red blood cells often exhibit axial migration, and leukocytes appear to flow primarily in the peripheral layer to the contrary, which is the first step in the firm adhesion to the endothelium [20]. In this study, the suspended particles were rigid and their shape and size were all the same. Judging from our results, it might be stated that margination occurs depending on concentration of the particles, when the particles are all rigid and spherical.

To consider the velocity profiles composed of the particles, the y-axis position and axial velocity of the particles were plotted in Fig. 3. The solid line indicates that from the Newtonian fluid. Our numerical data were in good agreement with that from the Newtonian parabolic fluid. This is because particles flowed in accordance with the streamlines of the macroscopic flows due to sufficiently weak inertial forces. The particles were scattered almost uniformly along the width (y-axis) direction, except for the peripheral layers near the wall ($y/l = \pm 1.0$) as mentioned above. However, some particles were pushed away toward the wall side and flowed within the peripheral layers for the case $\phi = 4.07\%$.

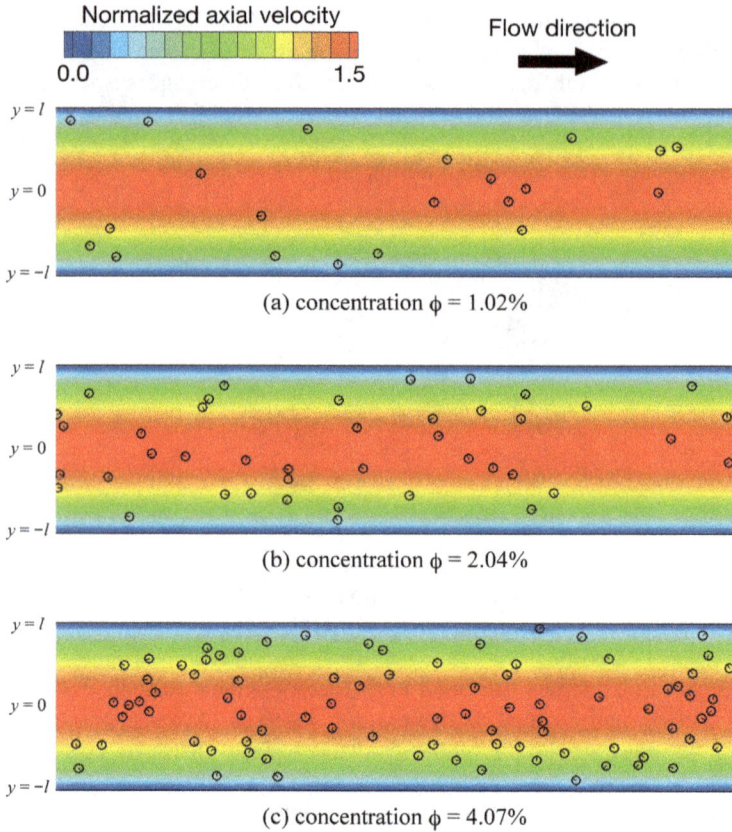

Figure 2:   Snapshots of pressure-driven suspension flows at time $t = 8$ s. (a) concentration $\phi = 1.02\%$; (b) concentration $\phi = 2.04\%$; and (c) concentration $\phi = 4.07\%$, respectively.

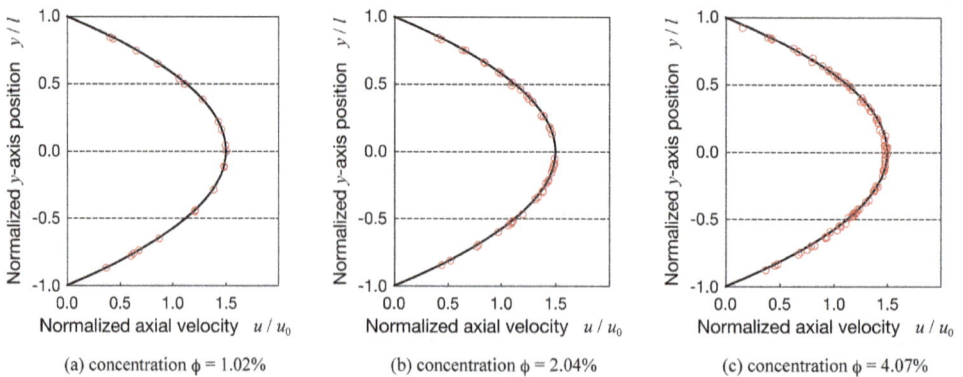

Figure 3:   Axial velocity profile composed of the suspended particles. The solid line indicates that from the Newtonian fluid: (a) concentration $\phi = 1.02\%$; (b) concentration $\phi = 2.04\%$; and (c) concentration $\phi = 4.07\%$, respectively.

Number of particles versus y-axis position was obtained as a function of probability density function (PDF) to consider their dispersed states. Relationship between PDF and y-axis position is also useful to consider its microstructure, i.e., spatial arrangement of the suspended particles. In this study, y-axis position was divided into 20 segments as shown in Fig. 4. Therefore, when the particles are homogeneously dispersed, the PDF value corresponds to $1/20 = 0.05$ as denoted by solid line in Fig. 4. Note that the value of 0.05 also corresponds to the confinement, i.e., ratio of particle size to the channel width. The data are plotted together with the standard error (SE) of the mean. Each SE was sufficiently small, which indicates number of trials in order to exclude initial position effects would be plenty. It was found that the PDF values were around 0.05 with some variations except for the peripheral layers ($y/l = \pm 1.0$), indicating the particles flowed homogeneously and were dispersed uniformly in the y-axis direction. This is because inertial effects of the particles were negligible for low Reynolds number condition as we discussed in previous studies [8], [15], [21]. It is worth mentioning that the values of PDF were almost zero, i.e., no particles were observed, in the peripheral layers for the case $\phi = 1.02$ and 2.04% due to strong repulsive forces from the channel wall. For the case $\phi = 4.07\%$, on the other hand, they were more flat around 0.05 including the peripheral layers as discussed above. Fig. 4 clearly shows differences in the dispersed states of the particles due to margination. Accordingly, microstructure of suspension can be easily visualized, compared and considered by using PDF.

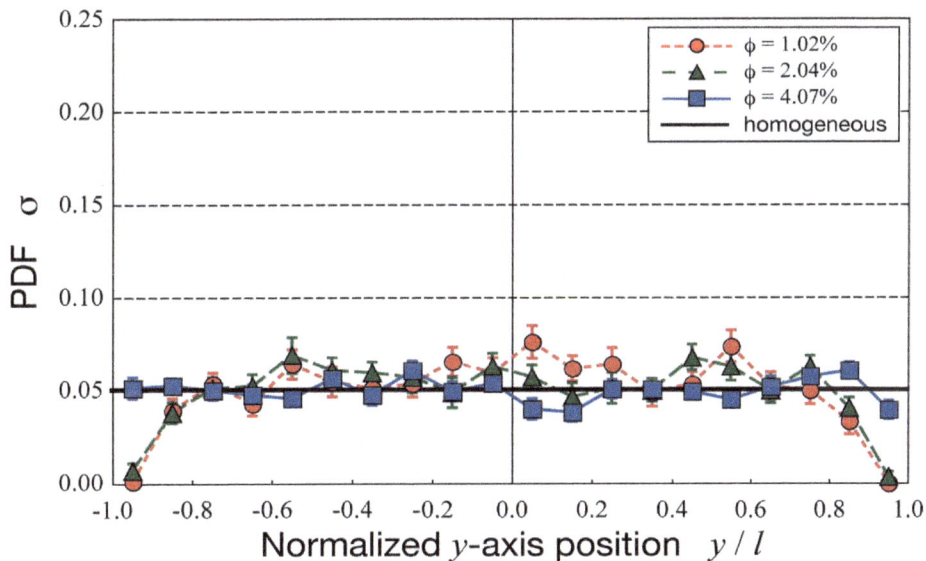

Figure 4:   Relationship between probability density function (PDF) and normalized y-axis position. The data are plotted together with ± 1 SE. The solid line denotes homogeneously dispersed state in the y axial direction.

Relative viscosity for each concentration is shown in Fig. 5. Data plotted are mean ± 1 standard error (SE). The error bars were sufficiently small, which indicates effects of initial particles' positions were properly removed. The solid line denotes that form Einstein's viscosity equation [2]. Our results were in good agreement with the theoretical values for lower concentration conditions, which indicates our computational conditions satisfied some assumptions in Einstein's equation, i.e., sufficiently small particles, negligibly weak inertial forces, homogeneously dispersed state, and low concentration. On the other hand, our data for $\phi = 4.07\%$ significantly differed from that by Einstein. Our data corresponded to that for $\phi = 6.8\%$ by Einstein's equation, which is 1.7 times as high as the actual concentration. We previously considered one of the reasons for nonlinearly enhanced relative viscosity for higher concentration suspensions from the viewpoint of rotational motions of the suspended particles due to hydrodynamic interactions [7]. In addition to this consideration, present study takes account of microstructure of the suspended particles. For higher concentration suspensions, particles flowed within the peripheral layers due to margination. When the macroscopic total effective viscosity can be discussed by summation of each microscopic particle's contribution to the effective viscosity, particles near the channel wall yield major and significant contributions [6], [14]. Then the intrinsic viscosity, which corresponds to a ratio of viscosity to concentration, was consequently increased with concentration. Microstructure using PDF would be a promising index to consider rheological properties of a suspension.

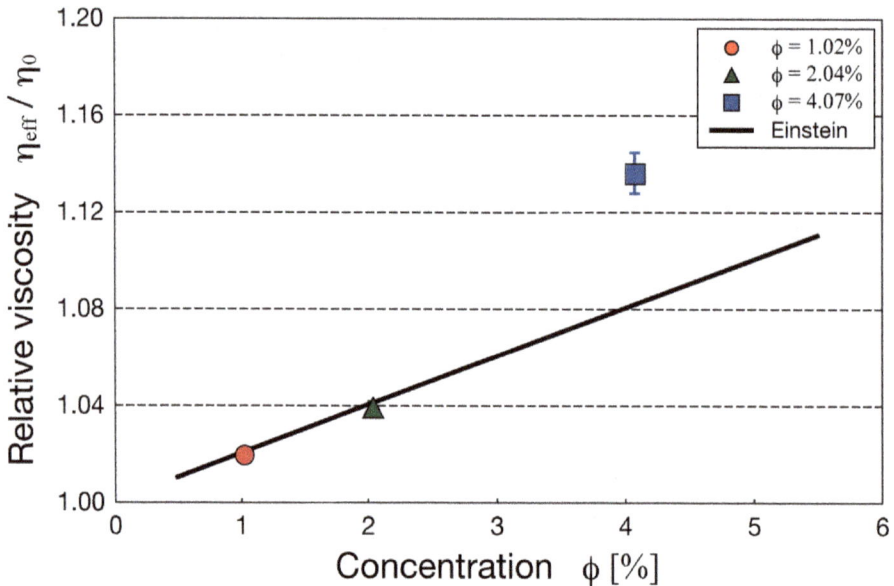

Figure 5:   Relationship between relative viscosity and concentration of a particle suspension. The data are plotted together with ±1 SE. The solid line denotes that from Einstein's viscosity equation.

4 CONCLUSIONS

Two-dimensional pressure-driven suspension flow simulations were conducted by a two-way coupling scheme in order to consider relationship between microstructure and consequent relative and intrinsic viscosities. As a result, for the case $\phi = 4.07\%$, the particles were dispersed homogeneously including peripheral layers due to margination. Then the relative and intrinsic viscosities were consequently enhanced and exceeded those from Einstein's estimation. Since the intrinsic viscosity was significantly related to the microstructure of the suspension, microstructure using PDF would be a promising index to consider rheological properties of a suspension.

ACKNOWLEDGEMENT

This work was supported in part by JSPS KAKENHI Grant Number JP20K04266.

REFERENCES

[1]   Einstein, A., Eine neue bestimmung der molekuldimensionen. *Annals of Physics*, **19**, pp. 289–306, 1906.

[2]   Brady, J.F., The Einstein viscosity correction in n dimensions. *International Journal of Multiphase Flow*, **10**, pp. 113–114, 1984.

[3]   Thomas, D.G., Transport characteristics of suspension: 8. A note on the viscosity of Newtonian suspensions of uniform spherical particles. *Journal of Colloid Science*, **20**, pp. 267–277, 1965.

[4]   Ghigliotti, G., Biben, T. & Misbah, C., Rheology of a dilute two-dimensional suspension of vesicles. *Journal of Fluid Mechanics*, **653**, pp. 489–518, 2010.

[5]   Mueller, S., Llewellin, E.W. & Mader, H.M., The rheology of suspension of solid particles. *Proceedings of the Royal Society A*, **466**, pp. 1201–1228, 2010.

[6]   Doyeux, V., Priem, S., Jibuti, L. & Farutin, A., Effective viscosity of two-dimensional suspensions: Confinement effects. *Physical Review Fluids*, **1**, 043301, pp. 1–22, 2016.

[7]   Fukui, T., Kawaguchi, M. & Morinishi, K., A two-way coupling scheme to model the effects of particle rotation on the rheological properties of a semidilute suspension. *Computers & Fluids*, **173**, pp. 6–16, 2018.

[8]   Fukui, T., Kawaguchi, M. & Morinishi, K., Numerical study on the inertial effects of particles on the rheology of a suspension. *Advances in Mechanical Engineering*, **11**, pp. 1–10, 2019.

[9]   Hu, X., Lin, J. & Ku, X., Inertial migration of circular particles in Poiseuille flow of a power-law fluid. *Physics of Fluids*, **31**, 073306, pp. 1–15, 2019.

[10]   Christ, F.E., Bowie, S. & Alexeev, A., Inertial migration of spherical particles in channel flow of power law fluids. *Physics of Fluids*, **32**, 083103, pp. 1–8, 2020.

[11]   Tanaka, M., Fukui, T., Kawaguchi, M. & Morinishi, K., Numerical simulation on the effects of power-law fluidic properties on the suspension rheology. *Journal of Fluid Science and Technology*, to be published.

[12]   Melander, O., Modrego, J., Zamorano-León, J.J., Santos-Sancho, J.M., Lahera, V. & López-Farré, A.J., New circulating biomarkers for predicting cardiovascular death in healthy population. *Journal of Cellular and Molecular Medicine*, **19**, pp. 2489–2499, 2015.

[13]   Stickel, J.J. & Powell, R.L., Fluid mechanics and rheology of dense suspensions. *Annual Review of Fluid Mechanics*, **37**, pp. 129–149, 2005.

[14]   Okamura, N., Fukui, T., Kawaguchi, M. & Morinishi, K., Influence of each cylinder's contribution on the total effective viscosity of a two-dimensional suspension by a two-way coupling scheme. *Journal of Fluid Science and Technology*, to be published.

[15]   Fukui, T., Kawaguchi, M. & Morinishi, K., Numerical study of the microstructure of a dilute suspension to assess its thixotropic behavior by a two-way coupling scheme. *WIT Transactions on Engineering Sciences*, vol. 128, WIT Press: Southampton and Boston, pp. 47–58, 2020.

[16] Morinishi, K. & Fukui, T., An Eulerian approach for fluid-structure interaction problems. *Computers & Fluids*, **65**, pp. 92–98, 2012.

[17] Izham, M., Fukui, T. & Morinishi, K., Application of regularized lattice Boltzmann method for incompressible flow simulation at high Reynolds number and flow with curved boundary. *Journal of Fluid Science and Technology*, **6**, pp. 812–821, 2011.

[18] Morinishi, K. & Fukui, T., Parallel computation of turbulent flows using moment base lattice Boltzmann method. *International Journal of Computational Fluid Dynamics*, **30**, pp. 363–369, 2016.

[19] Sterling, J.D. & Chen, S., Stability analysis of lattice Boltzmann methods. *Journal of Computational Physics*, **123**, pp. 196–206, 1996.

[20] Takeishi, N., Imai, Y., Nakaaki, K., Yamaguchi, T. & Ishikawa, T., Leukocyte margination at arteriole shear rate. *Physiological Reports*, **2**, pp. 1–8, e12037, 2014.

[21] Kawaguchi, M. et al., Viscosity estimation of a suspension with rigid spheres in circular microchannels using particle tracking velocimetry. *Micromachines*, **10**(10), 675, pp. 1–13, 2019.

SECTION 3
COMPUTATIONAL METHODS

SQUARE CYLINDER UNDER DIFFERENT TURBULENT INTENSITY CONDITIONS BY MEANS OF SMALL-SCALE TURBULENCE

ANTONIO J. ÁLVAREZ[1], FÉLIX NIETO[1], KENNY C. S. KWOK[2] & SANTIAGO HERNÁNDEZ[1]
[1]Structural Mechanics Research Group, University of La Coruña, Spain
[2]School of Civil Engineering, The University of Sydney, Australia

ABSTRACT

The phenomenon of turbulence is present in almost every type of flow in practical applications. Depending on its level of intensity and length scale, it can modify both the aerodynamic and aeroelastic performance of a body under flow action. In wind tunnel tests, the desired turbulence level is achieved by placing obstacles, spires, grids and extra roughness generators upwind the tested model. On the other hand, when trying to reproduce turbulence effects by means of a computational fluid dynamics (CFD) approach, two options have usually been considered: synthetic turbulence generation and the reproduction of velocity and pressure fluctuations recorded from previous simulations or wind tunnel tests. Another option, whose feasibility in CFD applications is addressed in this work by means of a 2D URANS (unsteady Reynolds averaged Navier–Stokes) consists of placing a rod upstream of the studied body, near the stagnation line. This approach is based on the generation of small scale turbulence upstream of the studied body, so that the turbulent wake generated by an upwind rod impinges on the body located downwind. In the present study, by means of 2D URANS simulations, the smooth flow over a circular cylinder (the upwind rod) is studied focusing on its wake turbulence characteristics. Furthermore, the aerodynamic performance of a square cylinder, first under smooth flow, and later immersed in the turbulent wake of the upstream rod, are analysed. A substantial effort has been devoted in the verification studies of the numerical models. It has been found that the adopted numerical approach is able to reproduce the turbulent characteristics of the rod wake and assess the impact of the turbulent flow on a square cylinder, providing a promising agreement with experimental data.
Keywords: 2D URANS, rod-generated turbulence, small scale turbulence (SST).

1 INTRODUCTION

Structures in the built environment are immersed in the atmospheric boundary layer, which is turbulent in nature. Therefore, the assessment of the aerodynamic and aeroelastic responses of structures encountering turbulent wind is of utmost importance to guarantee its safety and efficient performance.

The standard approach to study the effect of turbulence on a body is to conduct wind tunnel tests. In boundary layer wind tunnels, obstacles such as spires and roughness elements placed on the lower surface, are used to obtain the desired profile of mean velocity and turbulent intensity, representative of the turbulent atmospheric boundary layer. Alternatively, a grid placed upstream of the studied model may be used to generate a turbulent flow with the desired uniform mean speed and turbulent characteristics, without intending to replicate a boundary layer profile. Gartshore [1], proposed a different approach to generate small scale free stream turbulence by placing a rod along the stagnation line upstream of the studied body, to produce the major effects of free stream turbulence with the same turbulence intensity. This method, has been successfully applied, for instance, in Kwok and Melbourne [2] and Kwok [3]; and more recently in Lander et al. [4], to study the shear layer development of bluff bodies in a turbulent flow by means of Time Resolved Particle Image Velocimetry (TR-PIV).

Addressing free stream turbulence in CFD simulations represents a challenge due to the intrinsic unsteadiness, three dimensionality, broad range of scales and randomness in the incoming flow. According to Patruno and Ricci [5], there are two different methodologies to tackle this problem by adopting scale-resolving turbulence models such as Large Eddy Simulation (LES): one based on extracting the turbulent fluctuations from an auxiliary simulation or the recycling of the velocity in a plane within the simulation itself; and the generation of synthetic random fields. Both methods are complex and computational expensive, requiring a careful preparation [6], [7]. On the other hand, the use of URANS models to address free stream turbulence has been mainly confined to urban applications [8]–[10] due to its sufficient reliability for the intended applications and the much higher computational cost of scale-resolving alternative approaches.

This piece of research aims at assessing the feasibility of adopting a 2D URANS approach to study, at least qualitatively, the aerodynamic response of bluff bodies under free stream turbulent flow. To this end, the small-scale rod-generated turbulent flow approach proposed by Gartshore [1] for wind tunnel testing has been adopted in the computational simulations reported in this paper. A static square prism under smooth and 3.3% turbulent intensity (TI) free stream flows has been studied, comparing the numerical results with experimental data presented in Gartshore [1] and Lander et al. [4], as well as additional experimental data in the open literature. A good agreement has been found for important outputs such as the drag coefficient, pressure coefficient distributions and time-averaged streamlines. Hence, this preliminary study has shown promising results, suggesting undertaking further studies on this 2D URANS approach that decreases the computational burden of numerically studying turbulence effects while provides reliable results for wind engineering applications.

2 FORMULATION

2.1 Governing equations

The time averaging of the Navier–Stokes equations in a conservative form yields the URANS eqns [10]:

$$\frac{\partial U_i}{\partial x_i} = 0, - \tag{1}$$

$$\rho \frac{\partial U_i}{\partial t} + \rho U_j \frac{\partial U_i}{\partial x_j} = -\frac{\partial P}{\partial x_i} + \frac{\partial}{\partial x_j}(2\mu S_{ij} - \rho \overline{u_i' u_j'}) \tag{2}$$

where U_i is the mean velocity vector, x_i is the position vector, ρ is the fluid density, t is the time, P is the mean pressure, μ is the fluid viscosity, S_{ij} is the mean strain-rate tensor and u_i' is the fluctuating velocity.

The term $-\overline{u_i' u_j'}$ is the so-called specific Reynolds stress tensor (τ_{ij}), calculated by means of the Boussinesq assumption as:

$$\tau_{ij} = 2\nu_T S_{ij} - \frac{2}{3}k\delta_{ij}, \tag{3}$$

where ν_t is the kinematic eddy viscosity and k is the kinetic energy per unit mass of the turbulent fluctuation.

The different closure equations added to the previous ones, define the type of URANS model obtained. In this piece of research, the model selected is the $k - \omega$ SST, for incompressible flow, implemented in the open source software OpenFOAM, whose formulation was proposed by Menter and Esch [11].

2.2 Force coefficients, Strouhal number, pressure coefficient and base pressure coefficient

The time-dependent force coefficients (drag (C_d), lift (C_l) and moment (C_m)) along with the Strouhal (St) number, also referred to as integral parameters, are calculated according to eqn (4):

$$C_d = \frac{F_D}{\frac{1}{2}\rho U^2 D_s}, C_l = \frac{F_L}{\frac{1}{2}\rho U^2 D_s}, C_m = \frac{M}{\frac{1}{2}\rho U^2 D_s^2}, St = \frac{fD_s}{U}, \tag{4}$$

where D_s stands for the side of the square cylinder (see Fig. 1), ρ is the air density, U is the free-stream velocity, f is the dominant frequency of the lift coefficient, and F_D, F_L and M are the drag and lift forces and moment per unit of length respectively, which were calculated as the spanwise averaging of the integration of the pressure and viscous forces along the twin-box surfaces. The sign convention of the force coefficient is depicted in Fig. 1.

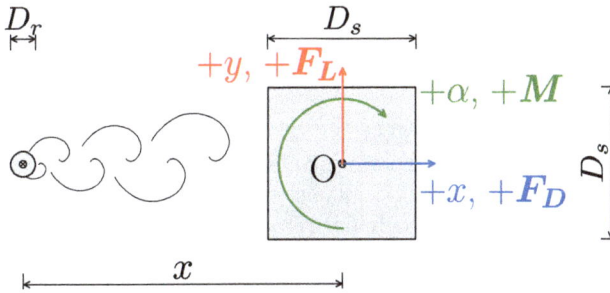

Figure 1:   Sign convention (D_s stands for the square cylinder side length and D_r for the rod diameter).

In the following, the time-averaged force coefficients values will be referred as $\overline{C_k}$ and their standard deviations as $\widetilde{C_k}$ $(k = d, l, m)$.

The mean pressure coefficient $\left(\overline{C_p}\right)$ and its standard deviation $\left(\widetilde{C_p}\right)$ are calculated as shown by eqn (5):

$$\overline{C_p} = \frac{\overline{p}}{\frac{1}{2}\rho U^2}, \widetilde{C_p} = \frac{\tilde{p}}{\frac{1}{2}\rho U^2}, \tag{5}$$

The base pressure coefficient $\left(C_{pb}\right)$ is calculated as indicated in eqn (6). The integral of the pressures is done over the complete leeward face of the square cylinder.

$$C_{pb} = \frac{\int_o^{D_s} \overline{p}(s)ds}{\frac{1}{2}\rho U^2 D_s}. \tag{6}$$

2.3 Turbulence intensity

According to Simiu and Scanlan [12], the longitudinal turbulence intensity of a wind flow (It_u) is:

$$It_u = \frac{\widetilde{U_x}}{U},$$

(7)

with $\widetilde{U_x}$ being the standard deviation of the longitudinal component of the wind. As the wind flow is obtained by means of URANS simulations, this value would only refer to the variation of the mean velocity, thereby missing the contribution of the fluctuating component. The fluctuating components are the ones yielding the specific Reynolds stress tensor, and according to Lander et al. [4] the turbulence intensity due to the fluctuating components of the longitudinal component of the velocity (It_R) is:

$$It_R = \frac{\sqrt{u_x' u_x'}}{U} = \frac{\sqrt{\tau_{xx}}}{U},$$

(8)

where τ_{xx} is the longitudinal component of the specific Reynolds stress tensor. Hence the total turbulence intensity (It) is the summation of both contributions:

$$It = It_u + It_R.$$

(9)

3 MODELLING AND COMPUTATIONAL APPROACH

Three types of simulations have been performed in this study: simulations of the isolated rod, simulations of the isolated square cylinder and simulations of the square cylinder in the wake of the rod. The side of the square cylinder (D_s) is twelve times the diameter of the rod ($D_r = D_s/12$), and all the simulations have been conducted at a Reynolds number, calculated with respect D_s, Re $= 3.84 \times 10^4$, the same used in the experiments conducted by Gartshore [1] that are later used for validation. These 2D URANS simulations were conducted by means of the $k - \omega$ SST turbulence model implemented in the CFD software OpenFOAM. The diffusive terms are computed using a second-order differential scheme, while the convective terms use the linear upwind differential scheme. The advancement in time is accomplish by a first order implicit scheme and the pressure velocity coupling is resolved by the PIMPLE algorithm.

The overall fluid domain dimensions, for the different types of simulations, is depicted in Fig. 2(a), and its dimensions are presented in Table 1.

For the space discretisation, a non-conformal structured quadrangular mesh is used. The fluid domain has been subdivided in five different zones (see Fig. 2(b)). In each boundary between zones, the number of elements is halved from the zone with a lower identifier. In all the simulations, the mean y^+ of the rod is always below 2.5, while this value is always below 1.2 for the square cylinder. In both cases the y^+ value has been calculated considering the total height of the first element of the boundary layer.

At the inlet, Neumann conditions were imposed for the pressure, while Dirichlet conditions were applied to the velocity, the specific dissipation rate and the turbulent kinetic energy. The last two values have been calculated considering an incoming turbulence intensity of 1.0% and a length scale of $0.1D_s$. In the case of the outlet boundary, Neumann conditions have been considered for the velocity, the specific dissipation rate and the turbulent kinetic energy fields, while Dirichlet conditions were applied to the pressure. For

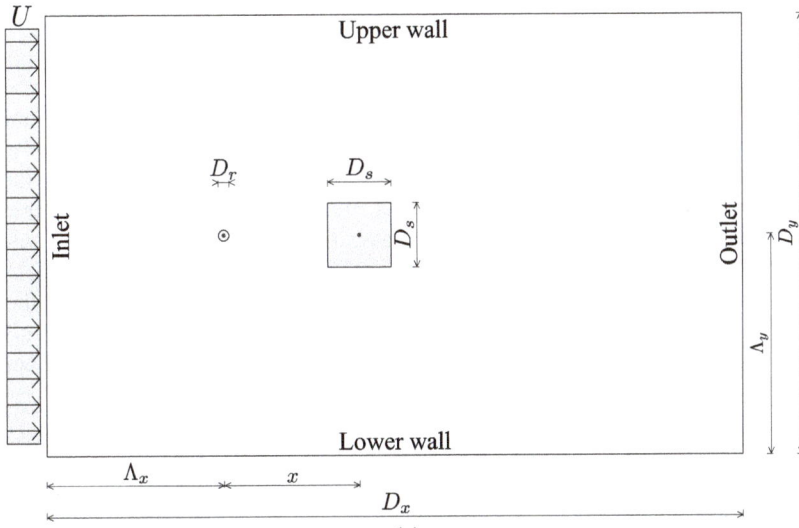

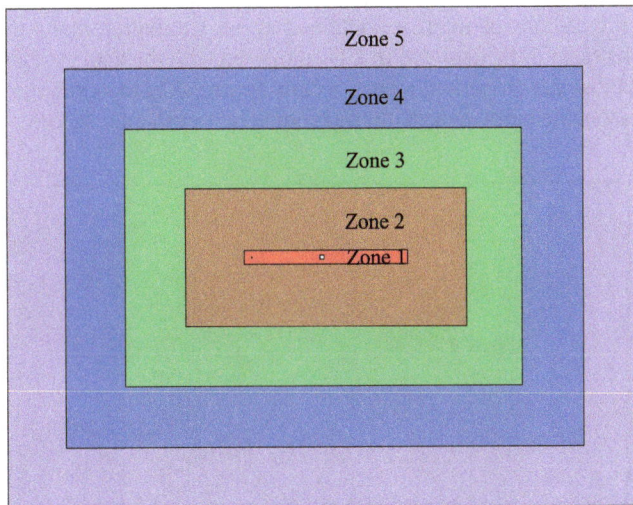

Figure 2:   Fluid domain definition. (a) Overall fluid domain (not to scale); (b) Different zones in which the mesh is subdivided.

Table 1:   Overall fluid domain dimensions. D_s is the side of the square cylinder, Λx is the distance from the centre of the rod to the inlet boundary, Λy is the distance from the centre of the square cylinder to both the upper and lower walls, Dx and Dy are respectively the total width and height of the fluid domain.

Λx	Λy	Dx	Dy
$60D_s$	$60D_s$	$160.5D_s + x$	$120D_s$

the upper and lower boundaries of the fluid domain a slip wall boundary condition was selected. In the deck walls, no penetration and no-slip boundary conditions were applied.

Verification studies aiming at identifying the finite volume grids insensitive to the spatial discretization have been conducted for the isolated rod and square cylinder in smooth flow. For the rod, three different grids were considered: a coarse mesh comprising 336,176 cells, a medium mesh comprising 507,984 cells and a fine mesh with 715,888 cells. It was found that the medium mesh provided results that were independent of the level of spatial discretization. For the isolated square prism, a similar study was carried out, also considering a coarse grid with 316,976 cells, a medium grid comprising 478,224 cells and a fine grid of 672,112 cells. In this case, the medium mesh grid also provided results that were insensitive to the spatial discretization level. The characteristics of the medium meshes were retained to generate the mesh combining the rod and the square prism, resulting in a 665,284 cells non-conformal structured hexahedral grid.

4 TURBULENCE GENERATION

Turbulence is generated by means of a circular rod placed upwind of the square cylinder, reproducing the arrangement in the experiments conducted by Gartshore [1] and Lander et al. [4]. Therefore the turbulent flow is due to the wake shed by the rod, which impinges upon the square cylinder. The turbulence intensity decays with the distance from the rod, as shown in Fig. 3, where the turbulence intensity obtained in our simulation along the centre line of the domain, downstream of the isolated rod, is compared with the data provided by Gartshore [1]. It is observed that the numerical results follow the trend of the experimental values and they present a reasonable agreement with the reported experimental values.

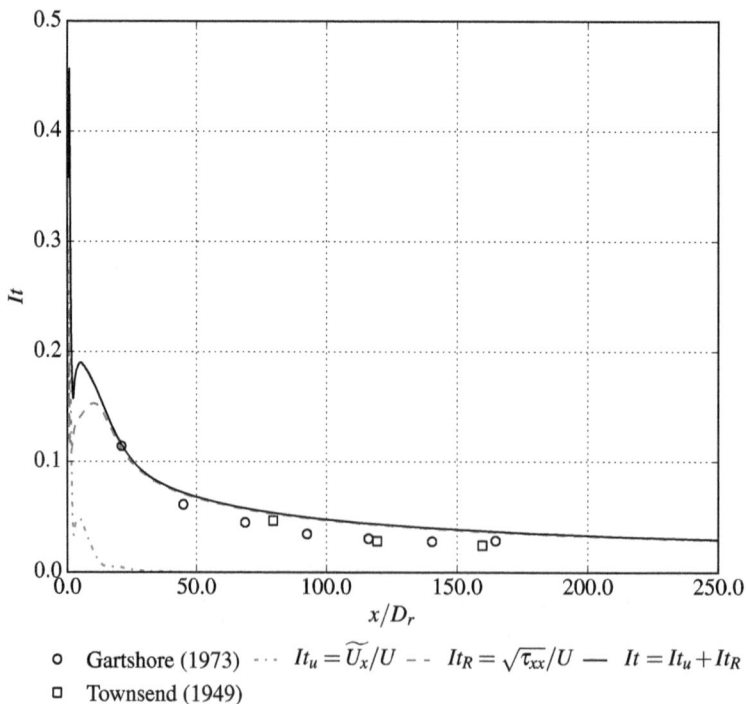

$$\circ \quad \text{Gartshore (1973)} \quad \cdots \quad It_u = \widetilde{U_x}/U \quad - - \quad It_R = \sqrt{\widetilde{\tau_{xx}}}/U \quad \text{---} \quad It = It_u + It_R$$
$$\square \quad \text{Townsend (1949)}$$

Figure 3: Longitudinal turbulence intensity in the wake of the circular rod.

5 RESULTS

It is recalled that the results reported next for the square cylinder in smooth flow are obtained by means of a 2D URANS simulation comprising only the square cylinder. For the case considering a nominal turbulent intensity of 3.3%, the computational model comprises the rod upstream of the square cylinder, separated a distance between centres $x = 212.37D_r$, as depicted in Fig. 1.

Fig. 4 presents the turbulence intensity profile in the wake of the rod along the stagnation line of and in the presence of the square cylinder. It is observed that the turbulence intensity diminishes as the distance from the rod increases, following the same trend as the one depicted in Fig. 3. The turbulent intensity starts to increase again in the vicinity of the square prism, reaching a maximum near the surface of the windward face.

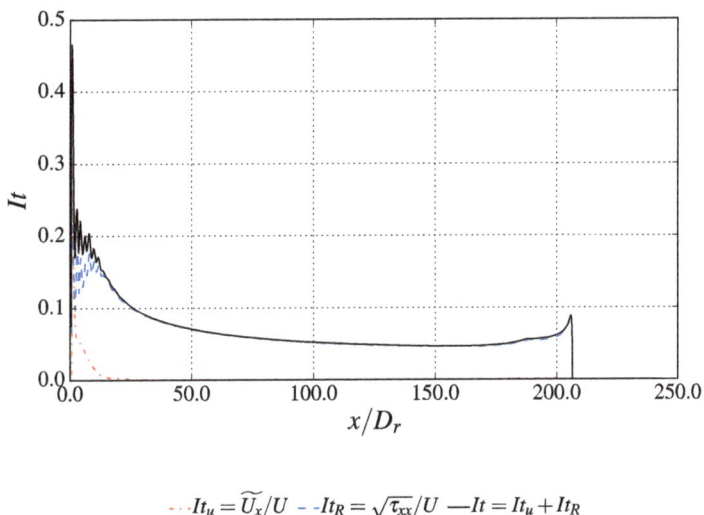

$$\cdots It_u = \widetilde{U}_x/U \quad ---It_R = \sqrt{\widetilde{\tau}_{xx}}/U \quad ---It = It_u + It_R$$

Figure 4: Longitudinal turbulence intensity in the gap between the rod and the square cylinder.

In Table 2, the force coefficients, Strouhal number and base pressure coefficient obtained for the square cylinder by the numerical simulations conducted in smooth flow are compared with available experimental results in the literature. Meanwhile in Table 3, the results for the same parameters are presented for the case considering a nominal turbulence intensity of 3.3%.

Table 2: Integral parameters and base pressure coefficient for the simulation in smooth flow. (PS refers to present study).

	Re	It %	$\overline{C_d}$	$\overline{C_l}$	$\overline{C_m}$	$\widetilde{C_d}$	$\widetilde{C_l}$	$\widetilde{C_m}$	St	C_{pb}
PS	3.84×10^4	0.0	2.06	0.02	-0.01	0.35	1.35	0.11	0.12	-1.27
[13]		0.0								-1.31
[14]	3.7×10^4	0.2	2.06	-0.02			1.02		0.12	-1.49
[1]	3.84×10^4	0.6	2.20							-1.45
[4]	5.0×10^4	1.0	2.35			0.22	1.14		0.13	-1.51

Table 3: Integral parameters and base pressure coefficient for the simulation under a total turbulence intensity (It) of 3.3%. (PS refers to present study).

	Re	$It\,\%$	$\overline{C_d}$	$\overline{C_l}$	$\overline{C_m}$	$\widetilde{C_d}$	$\widetilde{C_l}$	$\widetilde{C_m}$	St	C_{pb}
PS	3.84×10^4	3.3	1.92	-0.01	0.00	0.14	1.49	0.12	0.13	-1.26
[1]	3.84×10^4	3.3	1.84							-1.18
[4]	5.0×10^4	6.5	1.68			0.15	1.10		0.14	-1.22

Evidently, there are fairly good agreements between the numerical and experimental results, both in smooth and turbulent flows. Moreover, the drag coefficient mean value decreases as the level of turbulence intensity increases. On the other hand, the Strouhal number is slightly increased with an increase in turbulence intensity, for the experimental values.

In Fig. 5, the mean and fluctuating pressure coefficient distributions for smooth and 3.3% free stream turbulent flow are presented and compared with available experimental data.

The mean pressure coefficient distribution in smooth flow obtained by means of 2D URANS simulations agrees remarkably well with the experimental data in Carassale et al. [14]. For the numerical simulation considering a 3.3% turbulent intensity, a decrease in the pressure distribution on the windward face with respect to the smooth flow is obtained, which agrees with the trend reported in Lander et al. [4] for increasing turbulence levels. In the simulation, the mean pressure distribution on both side faces is relatively insensitive to the turbulence level, as in Lander et al. [4]. On the contrary, the decrease in the suction acting on the leeward face of the cylinder, identified experimentally in Lander et al. [4], is not evident in the numerical simulations, which generated a similar mean pressure distribution for both smooth and 3.3% turbulent flows. This circumstance will be further commented upon when referring to the mean streamlines plots.

With respect to the fluctuating component of the pressure coefficient, the numerical simulations not only reproduce the trends of the experimental values in Lander et al. [4], but are in a good agreement with the experimental data, remarkably in the reduction of the fluctuating pressure coefficient along the leeward face of the square cylinder with the increase in the free stream turbulence.

Finally, in Fig. 6, the time-averaged streamlines for smooth flow and 3.3% free stream turbulence obtained by 2D URANS are presented and compared with the experimental data for 1% and 6.5% free stream turbulence in Lander et al. [4]. The agreement between the experimental and numerical mean streamlines for nominal smooth flow may be highlighted. It has been possible to reproduce the size of the main vortex located in both side faces of the square cylinder, as well as the mean base-region size in its wake. On the other hand, when comparing the cases with higher free stream turbulence, the main vortices on both side faces of the square cylinder obtained by the 2D URANS simulation remain very similar to the ones in smooth flow, although a reduction in length in the recirculation region is observed due to the increase in curvature and the reattachment of the shear layer close to the leeward corners. This is qualitatively consistent with experimental evidence in Lander et al. [4]. However, the length of the base region for the 3.3% turbulent intensity case has not increased, as it should be expected.

These discrepancies may be explained by the limitations imposed on the turbulence model by the Boussinesq approximation, which concentrates the energy of the eddies in the vortex shedding frequency, to generate stronger vortices and therefore prevent the transfer of energy towards smaller scales of turbulence. Moreover, two-equation turbulence models tend to

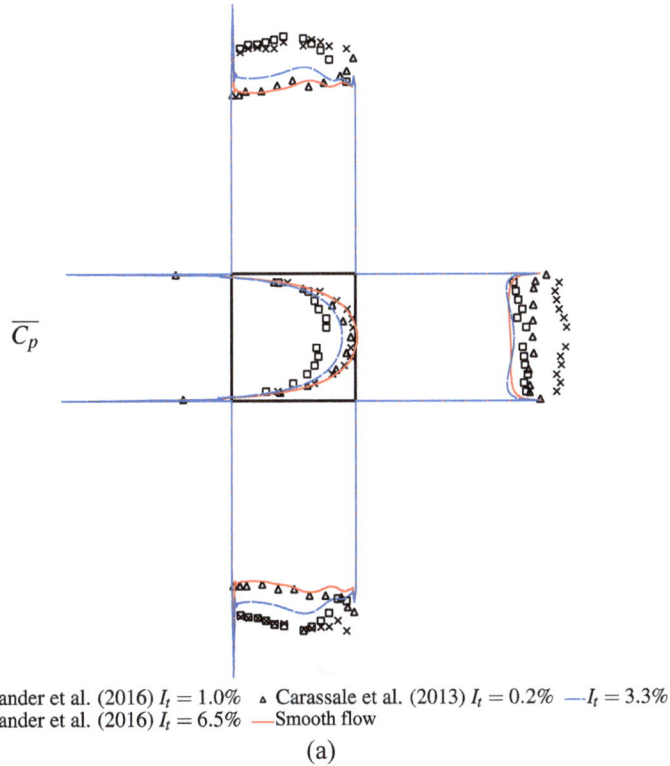

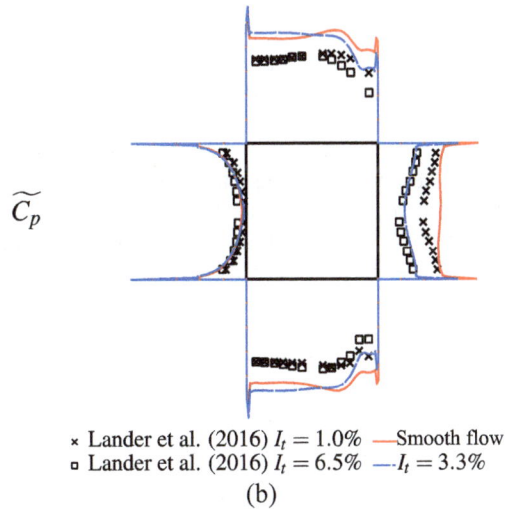

Figure 5:   Pressure coefficient distributions for the two levels of turbulence intensity considered. (a) Mean pressure coefficient distribution. Negative values points outward the section and the side of the square cylinder is equal to a value of unity; and (b) Fluctuating component of the pressure coefficient distribution. Positive values points outward the section and the side of the square cylinder is equal to a value of unity.

(a)

(b)

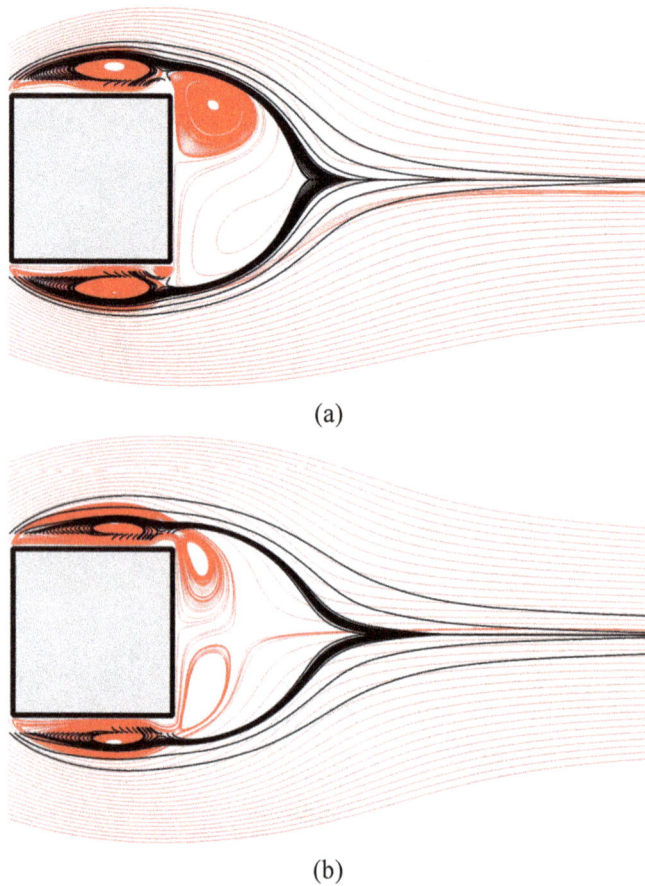

Figure 6: Time-averaged streamlines of the two cases with different turbulence intensity considered in the present study. The black lines correspond with the experimental data reported by Lander et al. [4] and the red lines are the ones corresponding to the present study. (a) Lander et al. 2016 It = 1.0% [4] vs present study smooth flow; and (b) Lander et al. 2016 It = 6.5% [4] vs present study It = 3.3%.

anticipate turbulent regimes at Reynolds numbers one order of magnitude lower than in experimental observations [15]. This prevents an accurate simulation of the transition phenomenon in the shear layer, which in this case could be further enhanced due to the addition of extra turbulent energy generated by the rod.

6 CONCLUSIONS

The main goal of this study has been to assess the feasibility of using relatively inexpensive numerical simulations that do not resolve the scales of turbulence to address the effects of free stream turbulence in bluff bodies. In the present study, 2D URANS simulations of a single rod, a single square and a combination of both have been carried out to study the effects of rod-generated small-scale turbulence on drag coefficient, Strouhal number, pressure coefficient distributions and time-averaged flow features.

The reported results have proven the ability of 2D URANS simulations to reproduce the effects caused by rod-generated uniform small-scale turbulence in the pressure distributions and mean flow features.

The proposed approach will form a suit of further study by completing a more systematic study for different free stream turbulence levels. Future work to be undertaken by the authors will comprise the consideration of different angles of attack and fluid-structure interaction problems such as lock-in and galloping under turbulent flow.

ACKNOWLEDGEMENTS

This research has been funded by the Galician Regional Government under the Competitive Reference Research Groups Program under the reference code ED431C2017/72. The computations have been carried out in the computer cluster Breogán and in the Galician Supercomputing Centre (CESGA).The authors gratefully acknowledge the support received.

REFERENCES

[1] Gartshore, I.S., The effects of free stream turbulence on the drag of rectangular two-dimensional prisms. Dept. BLWT-4-73, University of Western Ontario, Canada, 1973.

[2] Kwok, K.C.S. & Melbourne, W.H., Freestream turbulence effects on galloping. *ASCE Journal of the Engineering Mechanics Division*, **106**(2), pp. 273–288, 1980.

[3] Kwok, K.C.S., Turbulence effect on flow around circular cylinder. *ASCE Journal of the Engineering Mechanics Division*, **112**(11), pp. 1181–1197, 1986.

[4] Lander, D.C., Letchford, C.W., Amitay, M. & Kopp, G.A., Influence of the bluff body shear layers on the wake of a square prism in a turbulent flow. *Physical Review Fluids*, **1**, article no. 044406, 2016.

[5] Patruno, L. & Ricci, M., A systematic approach to the generation of synthetic turbulence using spectral methods. *Computer Methods in Applied Mechanics and Engineering*, **340**, pp. 881–904, 2018.

[6] Lamberti, G., García-Sánchez, C., Sousa, J. & Gorlé, C., Optimizing turbulent inflow conditions for large-eddy simulations of the atmospheric boundary layer. *Journal of Wind Engineering and Industrial Aerodynamics*, **177**, pp. 32–44, 2018.

[7] García-Sánchez, C., Van Tendeloo, G. & Gorlé, C., Quantifying inflow uncertainties in RANS simulations of urban pollutant dispersion. *Atmospheric Environment*, **161**, pp. 263–273, 2017.

[8] Antoniou, N., Motazeri, H., Neophytou, M. & Blocken, B., CFD simulation of urban microclimate: Validation using high-resolution field measurements. *Science of the Total Environment*, **695**, article no. 133743, 2019.

[9] Ricci, A. & Blocken, B., On the reliability of the 3D steady RANS approach in predicting microscale wind conditions in seaport areas: The case of the Ijmuiden sea lock. *Journal of Wind Engineering and Industrial Aerodynamics*, **207**, article no. 104437, 2020.

[10] Wilcox, D.C., *Turbulence modelling for CFD*, DCW Industries: La Cañada, 2006.

[11] Menter, F. & Esch, T., Elements of industrial heat transfer prediction. *Proceedings of the 16th Brazilian Congress of Mechanical Engineering*, **20**, pp. 117–127, 2001.

[12] Simiu, E. & Scanlan, R.H., Wind effects on structures, John Wiley & Sons: New York, 1986.

[13] Vickery, B.J., Fluctuating lift and drag on a long cylinder of square cross-section in a smooth and in a turbulent stream. *Journal of Fluid Mechanics*, **25**, pp. 481–494, 1966.

[14] Carassale, L., Freda, A. & Marrè-Brunenghi, M., Experimental investigation on the aerodynamic behavior of square cylinders with rounded corners. *Journal of Fluids and Structures*, **44**, pp. 195–204, 2014.

[15] Collie, S., Gerritsen, M. & Jackson, P., Performance of two-equation turbulence models for flat plate flows with leading edge bubbles. *Journal of Fluids and Structures*, **130**, pp. 021201-1-11, 2008.

[16] Townsend, A., Momentum and energy diffusion in the turbulent wake of a cylinder. *Proceedings of the Royal Society A*, **197**, pp. 124–140, 1949.

WIT Transactions on Engineering Sciences, Vol 132, © 2021 WIT Press
www.witpress.com, ISSN 1743-3533 (on-line)

VERIFICATION APPROACHES FOR THE 3D STATIC LES SIMULATIONS OF THE STONECUTTERS BRIDGE DECK

ANTONIO J. ÁLVAREZ, FÉLIX NIETO & SANTIAGO HERNÁNDEZ
University of La Coruña, Spain

ABSTRACT
In computational fluid dynamics (CFD) simulations, verification is the process of identifying both the spatial and the temporal discretisations providing relatively insensitive model results. The underlying goal of this task is to identify the mesh with a lower number of elements and the longer time step size compatible with the purposes of the study. In this manner, the computational burden associated with CFD methods may be decreased without compromising the accuracy of the numerical results. In general, this process is accomplished by studying three meshes with increasing discretisation levels (coarse, medium and fine), expecting that the medium one would provide satisfactory results. Once the satisfactory spatial discretisation level is selected, the different time step sizes are studied, taking into account that a Courant number (Co) of 1 is usually considered the maximum allowable value for numerical stability when adopting a LES (Large Eddy Simulation) approach. In this study, a detailed verification study considering two different approaches, providing the uncertainty level of several parameters of interest depending on the spatial and temporal discretisation is reported. The first of them consists in the curve fitting of linear or quadratic curves to selected model outputs; meanwhile the second approach relies on the Richardson extrapolation principle. These two different approaches are applied in the frame of the 3D LES numerical simulations of the Stonecutters bare deck geometry. In the study the focus is put on the sensitivity of the integral parameters, that is the force coefficients and Strouhal number, with the spatial and temporal discretisations.
Keywords: LES, stonecutters bridge, verification, uncertainty, force coefficients, Strouhal number.

1 INTRODUCTION
The terms of verification and validation must be used carefully as they refer to different specific processes, as it was stressed by Roache [1] in the context of computational fluid dynamics (CFD) computations. In this regard, verification consists of "solving the equations right", therefore it is a mathematical process, meanwhile validation refers to "solving the right equations", for which a sound scientific and/or engineering understanding of the phenomenon under study is needed, if the previous statement has to be fulfilled.

In bridge engineering applications it is common practice to carry out verification studies based on the evolution of certain parameters in three meshes with increasing grid refinement, aiming at reaching results not influenced by the level of discretisation. This procedure has been used in Álvarez et al. [2] and Laima et al. [3], although this kind of studies does not provide uncertainty levels of the parameters under study, which according to Oberkampf and Roy [4] is one of the four key factors to bring credibility and accuracy to the presented results, generating information of quality of a physical phenomenon, process or system.

According to American Society of Mechanical Engineers [5], verification is comprised of code verification and solution verification. The former is related to the accuracy in solving the mathematical model incorporated in the code, meanwhile the latter estimates the numerical accuracy of a particular calculation. Code verification is usually assumed [5], although as reported by Roache [1], it could be done by applying the method of manufactured solutions. Regarding solution verification, it is comprised by round off errors, due to the precision of computers, iterative errors, especially in time dependent solutions,

WIT Transactions on Engineering Sciences, Vol 132, © 2021 WIT Press
www.witpress.com, ISSN 1743-3533 (on-line)
doi:10.2495/MPF210101

and discretisation errors, due to the approximations made to transform the partial differential equations defining the flow into a system of algebraic equations [6]. Discretisation errors decrease as the grid resolution increases and, in general, it is the dominant source of uncertainty in practical CFD applications [6]. In this piece of research it is assumed that both the round-off and iterative errors are negligible, which in any case have to be two to three orders of magnitude lower than the discretisation error in order to guarantee a negligible influence [5].

The methods used in this paper, which to the author's knowledge is the first application of this methods to a bridge engineering application, are based on power series expansions, specifically the one depicted in Celik et al. [7] and the one presented by Eça and Hoekstra [6]. Therefore a quantifiable method is proposed for selecting the mesh with reasonable computational burdens and level of uncertainty.

2 FORMULATION

2.1 Governing equations

The movement of a fluid is defined by the Navier–Stokes equations. When they are modelled using a Large Eddy Simulation (LES) approach, the original equations are spatially filtered resulting in the following pair of equations [8]:

$$\frac{\partial \bar{u}_i}{\partial x_i} = 0, \tag{1}$$

$$\frac{\partial \bar{u}_i}{\partial t} + \frac{\partial \overline{u_i u_j}}{\partial x_j} = -\frac{1}{\rho}\frac{\partial \bar{p}}{\partial x_i} + \frac{\partial}{\partial x_j}\left[\nu\left(\frac{\partial \bar{u}_i}{\partial x_j} + \frac{\partial \bar{u}_j}{\partial x_i}\right) + \tau_{ij}^s\right], \tag{2}$$

where $\bar{u}$ is the filtered velocity, $\bar{p}$ is the filtered pressure, x is the space coordinate, t is the time, ν is the kinematic viscosity and ρ is the fluid density.

Using the Boussinesq assumption, the sub-grid stress tensor is expressed as:

$$\tau_{ij}^s = \nu_t\left(\frac{\partial \bar{u}_i}{\partial x_j} + \frac{\partial \bar{u}_j}{\partial x_i}\right), \tag{3}$$

where ν_t is the subgrid-scale turbulent viscosity.

The turbulence model selected for the present work is the Smagorinsky model [9]. This models is based on the assumption of equilibrium between the small resolved scales, dissipating the small ones all the energy extracted from the resolved ones. Moreover, it obeys the following equation:

$$\nu_t = (C_S\Delta)^2\left(2\overline{S_{ij}}\,\overline{S_{ij}}\right)^{1/2}, \tag{4}$$

where C_S is the Smagorinsky constant, $\overline{S_{ij}}$ is the filtered strain rate tensor and Δ is the characteristic spatial length of the filter, related to the mesh size, and defined as the cubic root of the mesh cell volume $\Delta = (\Delta V_i)^{1/3}$.

2.2 Force coefficients and Strouhal number

The time dependent force coefficients (drag (C_d), lift (C_l) and moment (C_m)) along with the Strouhal (St) number, also referred to as integral parameters, are calculated according to eqn (5):

$$C_d = \frac{F_D}{\frac{1}{2}\rho U^2 C}, C_l = \frac{F_L}{\frac{1}{2}\rho U^2 C}, C_m = \frac{M}{\frac{1}{2}\rho U^2 C^2}, St = \frac{fD}{U}, \tag{5}$$

where C stands for the width of a single box (see Fig. 1), D is the depth of a single box, ρ is the air density, U is the free-stream velocity, f is the dominant frequency of the lift coefficient, and F_D, F_L and M are the drag and lift forces and moment per unit of length, which were calculated as the spanwise averaging of the integration of the pressure and viscous forces along the twin-box surfaces. The sign convention of the force coefficient is depicted in Fig. 1.

In the following, the time averaged force coefficients values will be referred as $\overline{C_k}$ and their standard deviations as $\widetilde{C_k}$ ($k = d, l, m$).

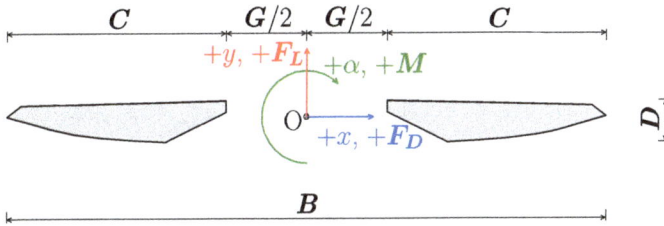

Figure 1: Sign convention.

2.3 Uncertainty calculation

For the calculation of the uncertainty due to the spatial discretisation of the integral parameters, the method described in Celik et al. [7], referred to here as "ASME method" and the one described in Eça and Hoekstra [6], named in the following as "Eça's method", are used. As previously commented, both methods are based on power expansion series, whose basic equation for estimation of the discretisation error (ϵ) is:

$$\epsilon_\phi \simeq \delta_{RE} = \phi_i - \phi_0 = \alpha h_i^p, \tag{6}$$

where ϕ_i is the parameter for which the uncertainty is to be calculated, ϕ_0 is the estimated exact solution, α is a constant, h is the representative cell size and p is the observed order of grid convergence.

According to Eça and Hoekstra [6] there are two assumptions that must hold for application of eqn (6): the grids are in the asymptotic range to guarantee that the leading term of the power series expansion is sufficient to estimate the error, and the level of refinement of the mesh can be represented by a single parameter, a representative cell size.

In both methods the representative cell size is calculated as:

$$h = \left[\frac{1}{N}\sum_{i=1}^{N}(\Delta V_i)\right]^{1/3}, \tag{7}$$

where N is the overall number of elements and ΔV is the cell volume.

2.3.1 ASME method

This method requires of three meshes with increasing refinement level, therefore $h_1 < h_2 < h_3$, and thus 1 refers to the fine mesh and 3 to the coarse one. The grid refinement factor $r = h_3/h_1$, has to be greater than 1.3 [7].

Eqns (8)–(11) show how the apparent order p of the method is calculated.

$$p = \frac{1}{\ln(r_{21})}|ln|\epsilon_{32}/\epsilon_{21}| + q(p)|, \tag{8}$$

$$q(p) = ln\left(\frac{r_{21}^p - s}{r_{32}^p - s}\right), \tag{9}$$

$$s = 1 \cdot sgn\left(\frac{\epsilon_{32}}{\epsilon_{21}}\right), \tag{10}$$

$$r_{21} = h_2/h_1, r_{32} = h_3/h_2, \epsilon_{21} = \phi_2 - \phi_1, \epsilon_{32} = \phi_3 - \phi_2. \tag{11}$$

Negative values of $\frac{\epsilon_{32}}{\epsilon_{21}} < 0$ are indicative of oscillatory convergence. The next step is to calculate the extrapolated values as:

$$\phi_{ext}^{21} = (l_{21}^p\phi_1 - \phi_2)/(l_{21}^p - 1), \phi_{ext}^{32} = (l_{32}^p\phi_2 - \phi_3)/(l_{32}^p - 1). \tag{12}$$

Moreover, along with the apparent order p, the approximate relative error (e_a see eqn (13)), the extrapolated relative error (e_{ext}, see eqn (14)) and the grid convergence index for the fine mesh (GCI_{fine}, see eqn (15)) has to be calculated and reported.

$$e_a^{21} = \left|\frac{\phi_1 - \phi_2}{\phi_1}\right|, \tag{13}$$

$$e_{ext}^{21} = \left|\frac{\phi_{ext}^{12} - \phi_1}{\phi_{ext}^{12}}\right|, \tag{14}$$

$$GCI_{fine}^{21} = \frac{1.25e_a^{21}}{r_{21}^p - 1}. \tag{15}$$

The uncertainty for each grid is calculated as shown in eqn (16). For further information about this method the interested reader is referred to Roache [1] and Celik et al. [7].

$$U_1 = \pm GCI_{fine}^{21}\phi_1. \tag{16}$$

2.3.2 Eça's method

In this method three alternatives equations are considered in eqn (6) for estimating the discretisation error:

$$\epsilon_\phi \simeq \delta_f = \phi_i - \phi_0 = \alpha h_i, \tag{17}$$

$$\epsilon_\phi \simeq \delta_s = \phi_i - \phi_0 = \alpha h_i^2, \tag{18}$$

$$\epsilon_\phi \simeq \delta_{fs} = \phi_i - \phi_0 = \alpha_f h_i + \alpha_s h_i^2. \tag{19}$$

To avoid unreliable results, at least four grids ($n_g \geq 4$) must be used to obtain a redundant system, which provides a quality check on the value of the apparent order of the method, and the unknowns in eqns (6) and (17)–(19) can be obtained by means of a least-squares approach. In this regard, the following equations should be minimised:

$$S_{RE}(\phi_0, \alpha, p) = \sqrt{\sum_{i=1}^{n_g} w_i \left(\phi_i - \left(\phi_0 + \alpha h_i^p \right) \right)^2}, \tag{20}$$

$$S_f(\phi_0, \alpha) = \sqrt{\sum_{i=1}^{n_g} w_i \left(\phi_i - \left(\phi_0 + \alpha h_i \right) \right)^2}, \tag{21}$$

$$S_s(\phi_0, \alpha) = \sqrt{\sum_{i=1}^{n_g} w_i \left(\phi_i - \left(\phi_0 + \alpha h_i^2 \right) \right)^2}, \tag{22}$$

$$S_{fs}(\phi_0, \alpha_f, \alpha_s) = \sqrt{\sum_{i=1}^{n_g} w_i \left(\phi_i - \left(\phi_0 + \alpha_f h_i + \alpha_s h_i^2 \right) \right)^2}. \tag{23}$$

With associated standard deviations:

$$\sigma_{RE} = \sqrt{\frac{\sum_{i=1}^{n_g} n w_i \left(\phi_i - \left(\phi_0 + \alpha h_i^p \right) \right)^2}{(n_g - 3)}}, \tag{24}$$

$$\sigma_f = \sqrt{\frac{\sum_{i=1}^{n_g} n w_i \left(\phi_i - \left(\phi_0 + \alpha h_i \right) \right)^2}{(n_g - 2)}}, \tag{25}$$

$$\sigma_s = \sqrt{\frac{\sum_{i=1}^{ng} nw_i \left(\phi_i - (\phi_0 + \alpha h_i^2) \right)^2}{(n_g - 2)}}, \tag{26}$$

$$\sigma_{fs} = \sqrt{\frac{\sum_{i=1}^{ng} nw_i \left(\phi_i - (\phi_0 + \alpha_f h_i + \alpha_s h_i^2) \right)^2}{(n_g - 3)}}. \tag{27}$$

Non-weighted and weighted approaches are used when performing the calculations. For the non-weighted approach,

$$w_i = 1, n = 1, \tag{28}$$

meanwhile for the weighted approach,

$$w_i = \frac{1/h_i}{\sum_{i=1}^{ng} 1/h_i}, n = n_g. \tag{29}$$

To calculate the discretisation uncertainty, first, eqn (6) is solved, for both the weighted and non-weighted approach. If the value of p of the two fits is $0.5 \leq p \leq 2$, the apparent order of the method is the one with the smallest standard deviation. If only one of the fits is inside the previous range, p is the one associated with that fit. Otherwise, if the observed apparent order of the method is $p > 2$, then eqns (17) and (18) are solved. In case of been $p < 0.5$, the eqns (17)–19 have to be solved. In all cases for both approaches, and selecting the apparent order of the method from the fit exhibiting the lower of the standard deviations.

Afterwards, the data range parameter, defined in eqn (30), is calculated in order to assess the quality of the fit.

$$\Delta_\phi = \frac{(\phi)_{max} - (\phi)_{min}}{n_g - 1}. \tag{30}$$

With this value and the apparent order of the method, the safety factor is calculated as:

$$\begin{cases} If\ 0.5 \leq p < 2.1\ and\ \sigma < \Delta_\phi, F_s = 1.25 \\ \quad\quad Otherwise, F_s = 3.0 \end{cases}. \tag{31}$$

Finally the uncertainty is calculated as:

$$\begin{cases} If\ \sigma < \Delta_\phi \Longrightarrow U_\phi(\phi_i) = \pm \left(F_s |\epsilon_\phi(\phi_i)| + \sigma + |\phi_i - \phi_{fit}| \right) \\ If\ \sigma \geq \Delta_\phi \Longrightarrow \pm U_\phi(\phi_i) = \pm 3 \frac{\sigma}{\Delta_\phi} \left(|\epsilon_\phi(\phi_i)| + \sigma + |\phi_i - \phi_{fit}| \right) \end{cases}. \tag{32}$$

For further information on the method the interested reader is referred to Eça and Hoekstra [6] and Rocha et al. [10].

Moreover, a slight modification of the previous model is proposed, which consists of setting the safety factor always equal to 1.25. This modification is based on three points:

- The recommendation of American Society of Mechanical Engineers [5] for using this less conservative value.
- The consideration of acceptable values of p outside of the range described in eqn (31) in Celik et al. [7].
- Eça's method was tested on simulations performed with a modification of the SIMPLE algorithm, which is thought for steady responses, and which is not the case for the application presented in this work.

3 MODELLING AND COMPUTATIONAL APPROACH

This study is performed over the geometry of the bare deck cross section of the Stonecutters bridge without modelling the transversal beams linking the boxes. The integral parameters are obtained by means of static 3D LES simulations, which were carried out using the CFD software OpenFOAM. The convective terms were discretised by using the second order upwind differencing scheme, while the second order central difference scheme was applied to the diffusive terms. The second order backward scheme was used for the advancement in time, and finally, the pressure-velocity coupling was solved by the PIMPLE algorithm.

The overall fluid domain is depicted in Fig. 2(a), their main dimensions are shown in Table 1. The spanwise dimension in this study is equal to the width of a single box.

At the inlet Dirichlet conditions were applied to the velocity and turbulent kinetic energy, meanwhile Neumann conditions were imposed to the pressure. At the outlet, Dirichlet conditions were applied to the pressure, and Neumann conditions to the velocity and turbulent kinetic energy. For the upper, lower and lateral faces, symmetric boundary conditions were applied. The incoming flow has a turbulence intensity of 0.0%. In the deck walls no penetration and no-slip boundary conditions were applied.

4 MESH CHARACTERISTICS

Two types of meshes were used for the discretisation of the fluid domain, a structured quadrangular one for the boundary layer and in the spanwise dimension, as the meshes were generated by extrusion, and an unstructured quadrangular mesh for the rest of the XY plane. This plane has been subdivided in different regions, whose outline is depicted in Fig. 2(b).

In Table 2, the non-dimensional sizes of the cells located in the boundaries of each zone are presented. Their size increase as they are placed further away from the deck. For Zone **L,** only the sizes of the elements located in its right bound are presented, as the size of the elements located in its upper and lower limits grow following a geometric series, starting from the size of Zone K, till reaching the one reported for Zone **L**. The elements of the right borders of Zone **M**, also increase following a geometric series, from the size of the elements in Zone **L** to the ones of Zone **M**. On the other hand, the characteristics of the boundary layer, the same for all the meshes considered herein, are presented on Table 3.

All the simulations were run at a $Re_D = 4.48 \cdot 10^5$ and Courant number $Co = 1$, presenting all the meshes a mean y^+ value, calculated as indicated in Bruno et al. [8], very close to 1.

(a)

(b)

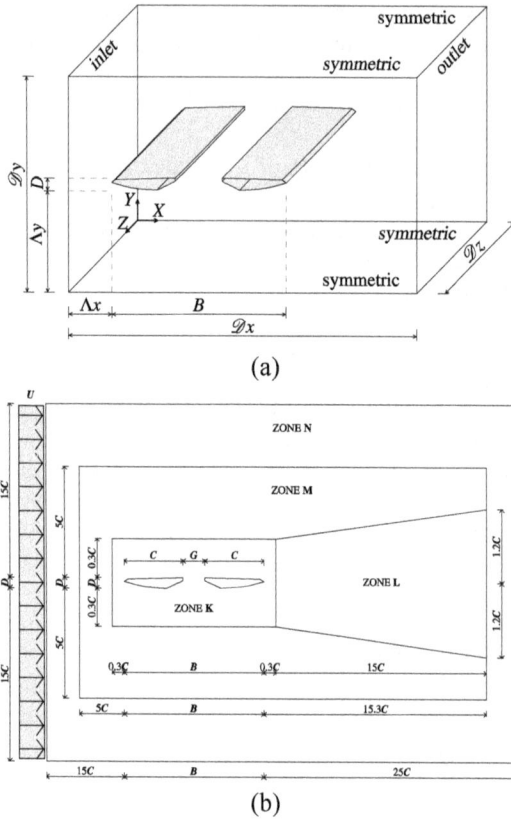

Figure 2:   (a) Overall fluid domain; and (b) Different zones in which the XY plane mesh is subdivided. (Not to scale.)

Table 1:   Overall fluid domain dimensions. B is the deck width, and C and D are the width and height of each individual box.

Λ_x	Λ_y	$\mathcal{D}_x$	$\mathcal{D}_y$	$\mathcal{D}_z$
$15C$	$15C$	$40C + B$	$30C + D$	C

Table 2:   Cell sizes on the borders of the different discretisation zones. The lower the number of the mesh, the higher the level of refinement is. All the values are non-dimensional, as the sizes have been divided by C. The computational cost was calculated over a dimensionless time unit ($t^* = tU/C = 1$).

Mesh	Zone K	Zone L	Zone M	Zone N	#elements	#of cores	Time per core (hours)
1	0.008	0.083	0.208	0.418	8598192	48	1.262
2	0.014	0.109	0.251	0.559	7121232	48	1.033
3	0.021	0.145	0.33	0.8	3218112	48	0.423
4	0.031	0.2	0.45	1.0	1970208	48	0.254

Table 3: Boundary layer mesh properties. The parameter y_1 is the height of the first element of the boundary layer (BL) mesh, C is the width of a single box, x_1 is the length of first element in the BL, r is the growth ratio of the elements in the BL, n_{BL} is the number of layers forming the BL mesh, y_{BL} is the total height of the BL mesh, δ_z is the length of the cell in the spanwise dimension and n_z is the number of elements in the spanwise dimension.

y_1/C	x_1/y_1	r	n_{BL}	y_{BL}/C	δ_z/C	n_z
0.0007	4	1.32	6	0.0088	0.021	48

5 RESULTS

In Table 4, the values of the integral parameters, obtained from the 3D LES simulations, for all the meshes considered are presented. The convergence criteria followed for the mean and standard deviations of the force coefficients, as well as the Strouhal number, is based on the residuals of the corresponding variables. The residuals for a generic q variable were calculated as $\varphi_{res} = |(q_n - q_{n-1})/q_n| \cdot 100$ (n is the number of sampling windows), and the simulations were extended until the residual value was lower than 5%, as indicated in Bruno et al. [8]. These residuals were obtained for increasing lengths of the sampling window T_n, with $T_0 = 50$ and $T_n = T_{n-1} + 50$ (in non dimensional time units $((tU)/D)$) (see Fig. 3). These values are the ones used for the calculation of the discretisation uncertainty. In Table 5, the results of applying the ASME method described in Celik et al. [7] are presented, as well as all the mandatory parameters when using this method. They have been calculated using meshes 1, 3 and 4.

Table 4: Integral parameters values for the meshes considered in this study.

Mesh	$\overline{C_d}$	$\overline{C_l}$	$\overline{C_m}$	$\widetilde{C_d}$	$\widetilde{C_l}$	$\widetilde{C_m}$	St
1	0.151	−0.263	0.226	0.023	0.166	0.076	0.247
2	0.153	−0.267	0.210	0.024	0.180	0.076	0.247
3	0.153	−0.260	0.206	0.027	0.202	0.090	0.244
4	0.154	−0.244	0.183	0.033	0.244	0.118	0.235

Continuing with the alternative method (Eça's method), the values of the discretisation uncertainty obtained are presented in Table 6, in this case the values of the integral parameters provided by the four meshes are used. It can be observed that the uncertainty values yielded by the modified version of the model $(U_i^{mod}, i = 1, ... ,4)$ are smaller than the ones presented by the original version of the method.

It is observed that the ASME method provides higher uncertainty values for coarser meshes, meanwhile this is not always the case for Eça's method. This can be explained by the fact that, in the calculation of the uncertainty shown in eqn (31), one of the elements is the difference between the value of the parameter obtained in the simulation and the fitted one. Moreover, this phenomenon could be explained by the fact of using only four meshes, although further investigation has to be conducted in this regard, and also the demanding computational needs of the 3D LES simulations, makes the increment of the number of simulations a cumbersome task.

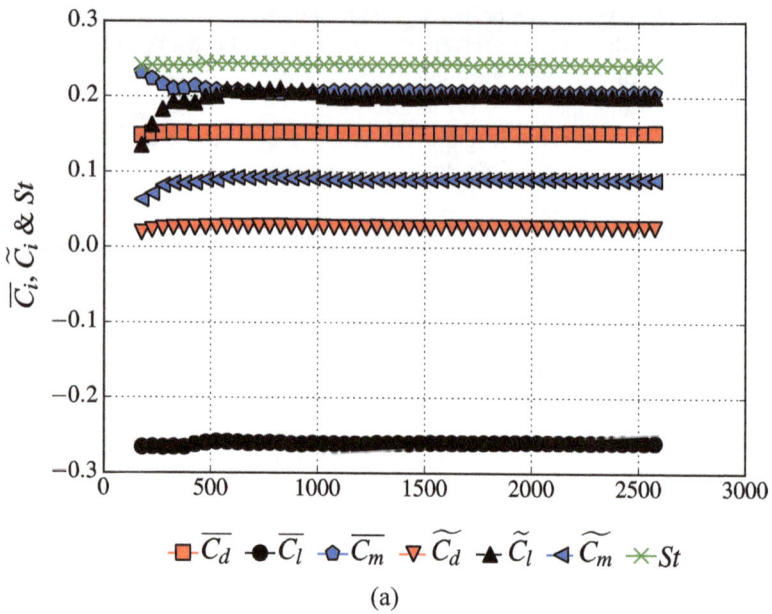

(a)

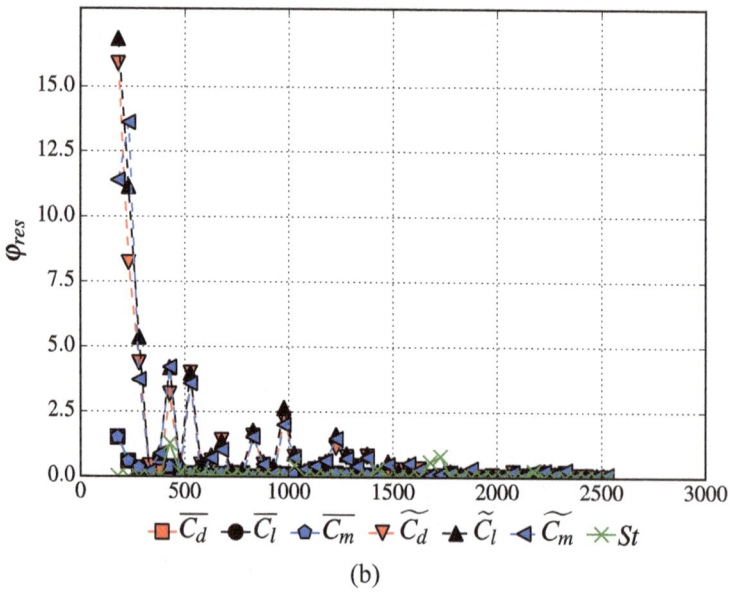

(b)

Figure 3:   (a) Integral parameters evolution; and (b) Integral parameters residuals for increasingly longer sampling windows, for mesh 3. Red color refers to statistical properties of the drag coefficient, black for the lift coefficient and blue for the moment coefficient, green color refers to the Strouhal number.

Table 5: Spatial discretisation uncertainty using the ASME method.

	$\overline{C_d}$	$\overline{C_l}$	$\overline{C_m}$	$\widetilde{C_d}$	$\widetilde{C_l}$	$\widetilde{C_m}$	St
N_1, N_3, N_4			8598192, 3218112, 1970208				
h_1, h_3, h_4			$4.881e^{-3}, 5.198e^{-3}, 6.773e^{-3}$				
r_{31}			1.388				
r_{43}			1.178				
ϕ_1	0.151	−0.263	0.226	0.023	0.166	0.076	0.247
ϕ_3	0.153	−0.260	0.206	0.027	0.202	0.090	0.244
ϕ_4	0.154	−0.244	0.183	0.033	0.244	0.118	0.235
p	2.990	11.661	3.507	4.653	3.588	6.190	7.858
ϕ_{ext}^{31}	0.150	−0.263	0.236	0.021	0.149	0.074	0.248
ϕ_{ext}^{43}	0.150	−0.263	0.235	0.021	0.149	0.074	0.248
e_a^{31}	1.138%	1.047%	8.999%	18.059%	21.959%	18.417%	1.333%
e_a^{43}	1.136%	6.202%	11.215%	22.301%	20.791%	31.380%	3.824%
e_{ext}^{31}	0.689%	0.023%	4.010%	5.295%	10.874%	2.872%	0.110%
e_{ext}^{43}	1.835%	1.070%	12.649%	24.311%	35.221%	21.819%	1.441%
GCI_{fine}^{31}	0.856%	0.029%	5.222%	6.286%	12.259%	3.492%	0.137%
GCI_{medium}^{43}	2.253%	1.352%	18.100%	24.446%	32.559%	22.389%	1.828%
GCI_{coarse}	9.781%	415.186%	101.314%	240.230%	189.642%	468.025%	86.679%
U_{fine}^{31}	$1.292e^{-3}$	$7.701e^{-5}$	$1.180e^{-2}$	$1.417e^{-3}$	$2.030e^{-2}$	$2.658e^{-3}$	$3.398e^{-4}$
U_{medium}^{43}	$3.441e^{-3}$	$3.512e^{-3}$	$3.722e^{-2}$	$6.506e^{-3}$	$6.576e^{-2}$	$2.019e^{-2}$	$4.459e^{-3}$
U_{coarse}	$1.511e^{-2}$	1.012	$1.849e^{-1}$	$7.819e^{-2}$	$4.627e^{-1}$	$5.546e^{-1}$	$2.034e^{-1}$
oscillatory convergence	✗	✗	✗	✗	✗	✗	✗

Table 6: Spatial discretisation uncertainty using Eça's method.

	$\overline{C_d}$	$\overline{C_l}$	$\overline{C_m}$	$\widetilde{C_d}$	$\widetilde{C_l}$	$\widetilde{C_m}$	St
ϕ_0	0.148	−0.278	0.242	0.017	0.125	0.005	0.255
$p_{observed}$	1.000	4.004	3.592	4.499	3.591	4.376	4.316
fit	eqn (17)	eqn (18)	eqn (18)	eqn (18)	eqn (18)	eqn (18)	eqn (18)
U_1	0.020	0.054	0.062	0.018	0.131	0.090	0.025
U_2	0.029	0.039	0.114	0.021	0.176	0.087	0.027
U_3	0.026	0.063	0.123	0.031	0.243	0.136	0.038
U_4	0.031	0.110	0.189	0.049	0.366	0.219	0.064
U_1^{mod}	0.020	0.027	0.033	0.008	0.060	0.042	0.012
U_2^{mod}	0.029	0.021	0.057	0.010	0.080	0.040	0.012
U_3^{mod}	0.026	0.032	0.059	0.014	0.109	0.063	0.018
U_4^{mod}	0.031	0.050	0.085	0.021	0.159	0.096	0.029
weighted	✗	✓	✗	✓	✓	✓	✓

6 CONCLUSIONS

In this piece of research the application of two different methods for the calculation of the discretisation uncertainty to the values obtained by means of 3D LES simulations have been presented for a practical case of interest in bridge engineering. The ASME method presented two main advantages respect to Eça's method: (i) it required a lower number of simulations and (ii) it has proved to be less conservative in the calculation of the uncertainty values. This suggests that the ASME method is more suitable for a wider range of practical applications. Moreover, the calculation of the uncertainties of the different parameters of interest provides major overall credibility to the obtained results, offering as well a quantifiable method for selecting a mesh with reasonable accuracy and computational demands, which could be used in further studies.

ACKNOWLEDGEMENTS

This research has been funded by the Spanish Ministry for Science and Innovation in the frame of the research project with reference PID2019-110786GB-I00 and the Galician regional Government with reference ED431C2017/72. The computations have been carried out in the computer cluster Breogan and in the Galician Supercomputing Center (CESGA). The authors fully acknowledge the received support.

REFERENCES

[1]   Roache, P.J., *Fundamentals of Verification and Validation*, Hermosa: Socorro, New Mexico, 2009.

[2]   Álvarez, A.J., Nieto, F., Kwok, K.C.S. & Hernández, S., A computational study on the aerodynamics of a twin-box bridge with a focus on the spanwise features. *Journal of Wind Engineering and Industrial Aerodynamics*, **209**, pp. 104465-1–18, 2021.

[3]   Laima, S., Jiang, C., Li, H., Chen, W. & Ou, J., A numerical investigation of Reynolds number sensitivity of flow characteristics around a twin-box girder. *Journal of Wind Engineering and Industrial Aerodynamics*, **172**, pp. 298–316, 2018.

[4]   Oberkampf, W.F. & Roy, C.J., *Verification and Validation in Scientific Computing*, Cambridge University Press: Cambridge, UK, 2010.

[5]   American Society of Mechanical Engineers, *Standard for Verification and Validation in Computational Fluid Dynamics and Heat Transfer*, ASME: New York, 2009.

[6]   Eça, L. & Hoekstra, M., A procedure for the estimation of the numerical uncertainty of CFD calculations based on grid refinement studies. *Journal of Computational Physics*, **262**, pp. 104–130, 2014.

[7]   Celik, I.B., Ghia, U., Roache, P.J., Freitas, C.J., Coleman, H. & Raad, P.E., Procedure for estimation and reporting of uncertainty due to discretization in CFD applications. *Journal of Fluids Engineering*, **130**, pp. 078001-1–4, 2008.

[8]   Bruno, L., Fransos, D., Coste, N. & Bosco, A., 3D flow around a rectangular cylinder: A computational study. *Journal of Wind Engineering and Industrial Aerodynamics*, **98**, pp. 263–276, 2010.

[9]   Smagorinsky, J., General circulation experiments with the primitives equations. I: The basic experiments. *Month. Weath. Rev.*, **3**(91), pp. 99–165, 1963.

[10] Rocha, A.L., Eça, L. & Vaz, G., On the numerical convergence properties of the calculation of the flow around the KVLCC2 tanker in unstructured grids. *Proceedings of the VII International Conference on Computational Methods in Marine Engineering*, pp. 336–352, 2017.

SECTION 4
THEORETICAL AND
COMPUTATIONAL
FORMULATIONS

STUDY ON PINHOLE LEAKS IN GAS PIPELINES: CFD SIMULATION AND ITS VALIDATION

BURAK AYYILDIZ[1], MUHAMMAD AZIZUR RAHMAN[1,2], ADOLFO DELGADO[1],
IBRAHIM HASSAN[2], HAZEM NOUNOU[2], RASHID HASSAN[1] & MOHAMED NOUNOU[2]
[1]Texas A&M University, USA
[2]Texas A&M University at Qatar, Qatar

ABSTRACT

In the present study, the computational fluid dynamics (CFD) simulations of pinhole leaks (1.27–3.3 mm) in a low-pressure, up to 2.5 bars, air pipeline which has 16 mm (0.62 inch) inner diameter has been performed by using a 3D transient DES (detachable eddy simulation) model of a commercial CFD code, ANSYS Fluent R3. Also, a laboratory-scale experimental setup is established with a 5 m long pipe with inner diameter (ID) = 16 mm. In steady operational stages, mass balance method is used to calculate the leakage mass flow rate in the experimental setup. In addition, pressure point analysis (PPA) with two dynamic and one differential pressure gauges is used to detect chronic/small leaks at transient stages. This method is cost effective and easy to maintain compared to expensive acoustic leak detection systems. The numerical results were validated against the experimental data. The simulations values of leakage mass flow rate are slightly higher (~10%) than the experiments but overall the simulation results are in good agreement with experiment. The proposed model simulates the flow leakage, pressure distribution and velocity profile around the defined size of the leakage. Transient simulation is performed to use power spectral density (PSD) and Fast Fourier transformation (FFT) of the acoustic pressure variation to predict acoustic oscillations and turbulent behavior of the flow field around the leakage location. These results could help advance current understandings of several leak detection systems that will reduce the false alarms of the leakage monitoring systems.
Keywords: CFD simulation, chronic leak detection, gas pipeline leakage, dynamic pressure wave monitoring.

1 INTRODUCTION

Pipelines have been used to transport water, fossil fuels, gases and chemicals, which are crucial for modern society, over millions of miles all around the world. Especially, oil and gas producer countries' (USA, Russia, UAE, Canada, Qatar, etc.) pipelines are vital link connecting reservoirs which are far away from main land to refineries, plants and consumers in the home/exported country. There are several ways to transfer petroleum products rather than pipeline, for instance rail or truck on land and oil tanker at off-shore, but using pipeline is more economical when you need to transport high volume of product over long distance.

The accidental release or admission of fluid through a hole is characterized as leak. There are several reasons of the leakage: bad workmanship, corrosion, excess fluid pressure change, cracks/defects in the pipeline, lack of maintenance and natural disasters.

Chronic leak detection is an important concern as they have the potential to be a significant contributor of green-house gases (GHGs) if they got undetected for extended periods of time [1]. Early and accurate detection is essential to minimize pipeline fracture propagation, continuous chronic leaks, and possible safety-related incidents.

Most industries collect abundant measurements from a variety of sensors monitoring different process variables, making it easy to draw relationships between different physical properties in order to carry out efficient fault detection, i.e., anomaly (or leak) detection [2]. Unfortunately, for subsea pipelines experiencing harsh conditions a limited number of

WIT Transactions on Engineering Sciences, Vol 132, © 2021 WIT Press
www.witpress.com, ISSN 1743-3533 (on-line)
doi:10.2495/MPF210111

sensors can be installed, due to installation and maintenance costs. Therefore, due to the lack of measurements from a multitude of different sensors, available sensor measurements need to be analyzed efficiently in order to carry out leak detection.

Ben-Mansour [3] performed 3D steady and transient turbulent CFD simulations for small leaks (below 1 l/min) in water pipelines. The results indicate clear influence of the leak in the pressure gradient. Also, there is a significant change at the magnitude of the acoustic signal in presence of the leak. Moreover, Ben-Mansour [4] developed a CFD model that shows axial flow acceleration in the flow mid-plane of the pipeline that indicates clear gradient jump which can be used as leak detection method.

Liu [5] proposed a wavelet transform (WT) method by using quadrupole and dipole sonic sources. However, the time difference (TD) error were large due to fact the leakage time was not clear. Liu [6] studied an improved dynamic pressure waves method that localization of the leak in natural gas pipeline. We have performed numerical simulations and experiments for single phase flow [7]–[9], gas/liquid flow [10]–[14], solid/liquid [15], gas/liquid/solid flow [16], [17]. We have also previously conducted a critical literature review on onshore/offshore leak detection and single phase (crude oil and natural gas) leak detection computational fluid dynamics modeling [18]. In this study, we have applied the wavelet transform method on the dataset that was obtained from our computational fluid dynamics modelling. We have found that wavelet transform method is very effective tool in identifying the leak detection and localization.

2 EXPERIMENTAL SETUP OF PINHOLE LEAK DETECTION FACILITY

A low-pressure pipeline gas(air) leak detection setup is designed and established by our group. The pressure drop and leak flow rate are measured across a length of 1-meter pipe section (distance between P1 and P2 in Fig. 1). The total length of the setup is 4 meters straight stainless-steel pipe of 0.75-inch outer diameter, 0.62-inch inner diameter. P&ID diagram of the single-phase leak detection test rig is shown in Fig. 1. The test rig is mounted on the top of lab bench shelves. The main air supply is connected to the university's air supply domain and provide air pressure up to 3.5 bars. There are two

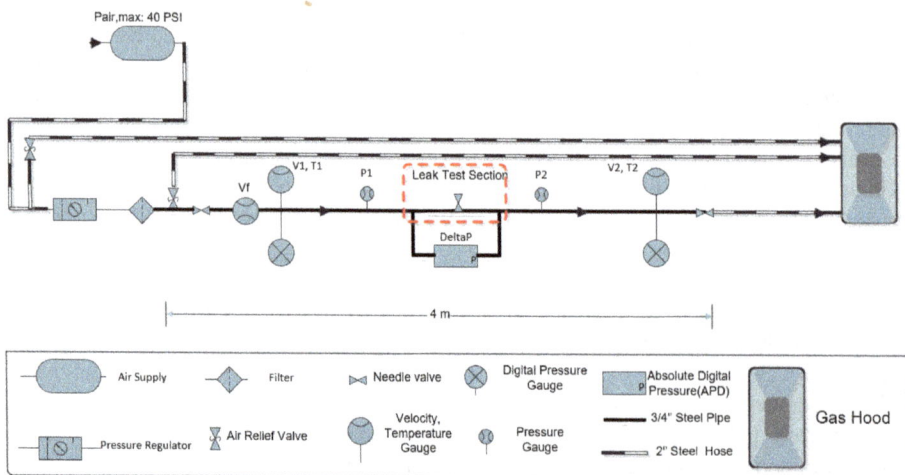

Figure 1:   P&ID of gas pipeline leak detection setup, the main instruments in the flow loop includes needle valve, solenoid valve, dynamic and differential pressure gauges.

pressure relief valves set to 4 bars pressure to maintain the safety of the main pipeline. Furthermore, pressure regulator and air filter are mounted at the beginning of the loop to provide dry and clean air to the system. The leak flow rate is controlled by various size of solid-stream spray air nozzles which are controlled by on/off solenoid valve. The diameter of the spray nozzles are 0.05″ (1.27 mm), 0.07″ (1.778 mm), 0.09″ (2.286 mm), 0.1″ (2.54 mm), and 0.13″ (3.302 mm).

Data acquired by sensor with a constant pressure with no leak condition, then open the solenoid valve to create leakage. The spray nozzle can be changed with a different nozzle size at the end of data collection. The main instruments are used in this flow loop are: ESI-dynamic pressure transducer (UPS-HSR-B02P5-N), Omega-flow and temperature sensor (FMA1003) and Omega-differential pressure sensor (PXM409-070HDWUUSBH).

3 TRANSIENT CFD SIMULATIONS AND FFT ANALYSIS

3.1 Mathematical model of leak simulation

3.1.1 Mass conservation
The equation of continuity (mass conservation) for transient flow can be written as eqn (1)

$$\frac{\partial \rho}{\partial t} + \nabla.\left(\rho \vec{v}\right) = 0 \,. \tag{1}$$

3.1.2 Momentum conservation
A single set of momentum equation is solved throughout the computational domain, and the resulting velocity field is shared by the volume fractures of fluids. The momentum eqn (2), depends on the phases' volume fractures through the density and viscosity properties.

$$\frac{\partial}{\partial t}\left(\rho \vec{v}\right) + \nabla.\left(\rho \vec{v}\vec{v}\right) = -\nabla p + \nabla.\left[\mu\left(\nabla \vec{v} + \nabla \vec{v}^{T}\right)\right] + \rho \vec{g} + \vec{F} \,. \tag{2}$$

3.1.3 Turbulence model
The DDES (Delayed Detached Eddy Simulation) is applied to simulation, in order to study the transient turbulent behavior of the flow field inside the pipeline at single phase flow conditions.

In the DES model, which is a hybrid turbulence model, the unsteady RANS (Reynolds-averaged Navier–Stokes) models are applied in the boundary layer while LES (Large Eddy Simulation) treatment is employed to the separated regions. The main advantage of DES over LES, DES model has been designed to address high Re number wall bounded flows, where it requires a smaller number of nodes. This model cost less computational time with its simplicity, respect to LES. However, the danger of grid-induced separation (GIS) where the flow separation depends on the critical value of h_{max}/δ where h_{max} is the max cell edge length and δ is the local boundary layer thickness, not the physics of the flow [19]. Therefore, DDES shielding function is introduced [20], [21]. The DDES shielding function aims to preserve the eddy viscosity from degradation up to $h_{max}/\delta = 0.1$. The empirical delay function is presented in eqn (3)

$$f_d = 1 - tanh\left[\left(C_{d1}r_d\right)^{C_{d2}}\right], \tag{3}$$

where C_{d1}, C_{d2} are 20, 3 respectively, and r_d shows in eqn (4)

WIT Transactions on Engineering Sciences, Vol 132, © 2021 WIT Press
www.witpress.com, ISSN 1743-3533 (on-line)

$$r_d = \frac{\upsilon_t + \upsilon}{\kappa^2 d_w^2 \sqrt{0.5\left(S^2 + \Omega^2\right)}} , \qquad (4)$$

where

υ_t : eddy viscosity

υ : molecular viscosity

S: strain rate

Ω : vorticity tensor

κ : von Karman constant (0.41)

d_w: distance to the wall

3.2 Solution procedure and mesh quality

A 3D turbulent flow simulation is conducted within the fluid domain of a pipe diameter D = 0.016 m and length of L = 2 m. The location of leak is at the center point of the mid-top section of the pipeline, shown in Fig. 2. The hole has a rectangular shape and to get a better resolution 10 cells per gap is defined. PISO (pressure implicit with splitting of operation) algorithm is used for pressure-velocity coupling scheme controls, because PISO may provide faster solution time at transient simulations for small time step sizes [22]. Table 1 indicated the details of boundary conditions and physical models.

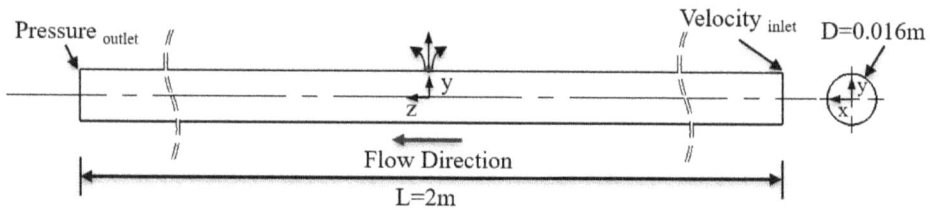

Figure 2:   Geometrical design of the leak detection simulation's model (Leak center at, x = 0, y = D/2 and z = 0).

Table 1: Details of boundary conditions and physical models.

Solution type	Transient-state (time-step: 0.001 s)
Solver type	Pressure-based
Fluid	Air (real-gas (peng-robertson)) Methane (real-gas (peng-robertson))
Inlet boundary conditions	Velocity inlet
Outlet boundary conditions	Pressure outlet
Turbulence model	DDES
Acoustic model	FW-H equation

To ensure number of cells is not affect the solution independency, the mesh independence study is performed. Five meshes were designed and used with the boundary conditions of pressure outlet 78,000 Pascal, velocity inlet as 4.2 m/s (Fig. 3). All meshes are maintained the lowest requirement of the acceptable mesh quality (minimum orthogonal quality >0.01 and maximum aspect ratio <40 for ANSYS Fluent). Due to fact we are using the supercomputer in TAMUQ and there is no cost constraints in this case, we decide to use 913 thousand polyhedral cells mesh model for our simulations to get a better accuracy and the resolution.

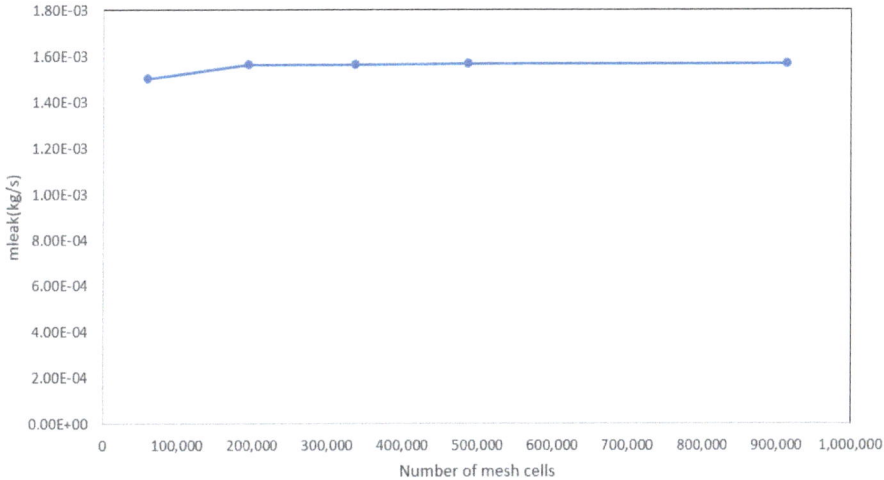

Figure 3: Mass leak flow rate versus number of cells.

3.3 Boundary conditions

The domain is named as velocity inlet, pressure outlet, leak pressure outlet and wall. The pipe inlet velocity and the pressure outlet values are prescribed from the experimental data set (Table 2). At the leak location, pressure outlet which is equal to the atmospheric condition is defined as boundary condition. Various leak sized (1.27, 1.788 and 3.302 mm) are also defined at simulations.

Table 2: Test cases and the conditions are measured from the experimental setup and used as boundary conditions in the simulations.

Case #	Inlet velocity (m/s)	Inlet temperature (°C)	Outlet pressure (bar)	Outlet temperature (°C)	Leak size (mm)
1	37.56	22.5	2.07	24.1	1.27
2	42.45	22.6	0.64	24.3	3.302
3	42.3	22.2	0.77	23.08	1.788
4	42.15	22.3	1.17	23.9	1.788

The numerical leakage flow rate calculations are validated with experimental results for air cases and later on, same model is modified by the methane within the simulation to perform industrial conditions.

3.4 Model validation

The simulation was carried out with different working pressure range between 0.2–2.5 bars and was validated with the experimental data. Fig. 4 shows the summary of the experimental and prediction. The simulations values of leakage mass flow rate are slightly higher (~10%) than the experiments but overall the simulation results are in good agreement with experiment. As the pressure inside the pipeline increases, pressure differences at the leak surface increases as well and cause higher leak flow rates. Moreover, leak flow rate increases as leak size increases, due to fact pressure loss across the leak surface decreases.

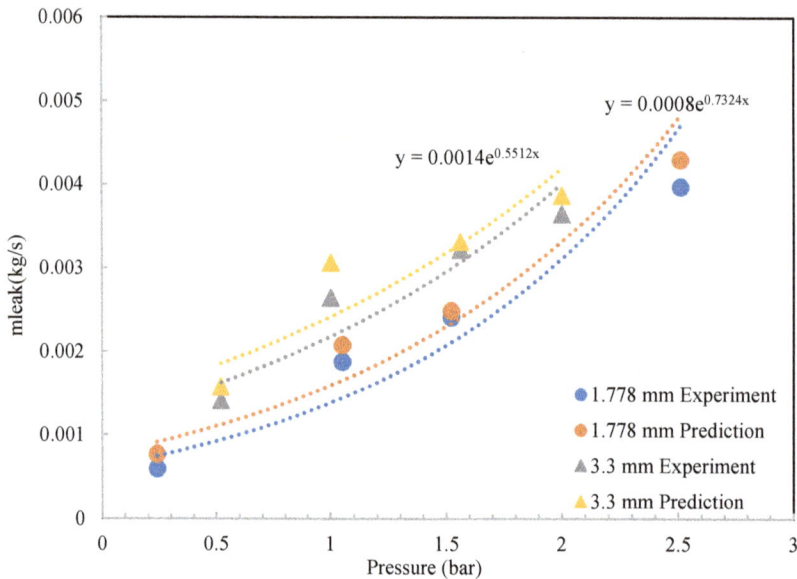

Figure 4: CFD prediction validation with experimental data on leak flow rate.

3.5 Pressure and flow behaviors

In this section, we study the pressure and flow behavior of air in the vicinity of a pinhole. Mainly case 3 is chosen to perform analysis at instantaneous time which is 10 seconds but all other cases have the similar trends. Fig. 5 illustrates the pressure contour around the 1.788 mm by 1.788 mm leak size. There is large pressure variation around the leak location, however the main pressure is dominating around the rest of the pipeline. The pressure is dropped to the atmospheric pressure gradually. This drop is limited with 1 mm region.

Fig. 6 indicates x-vorticity contour field which shows the circulation region around the leak location. There are two local circulation regions, the maximum x-vorticity after the leak region and minimum one before the leak region.

Fig. 7 indicates the temperature variation around the vicinity. Due to the local cooling effect can be explained by the Joules-Thompson effect during gas decompression,

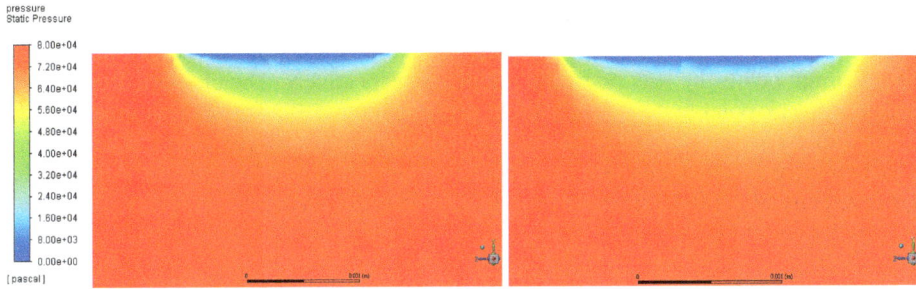

Figure 5:   Pressure contours around the leak position, case 3, time = 10 s (left: air; right: methane).

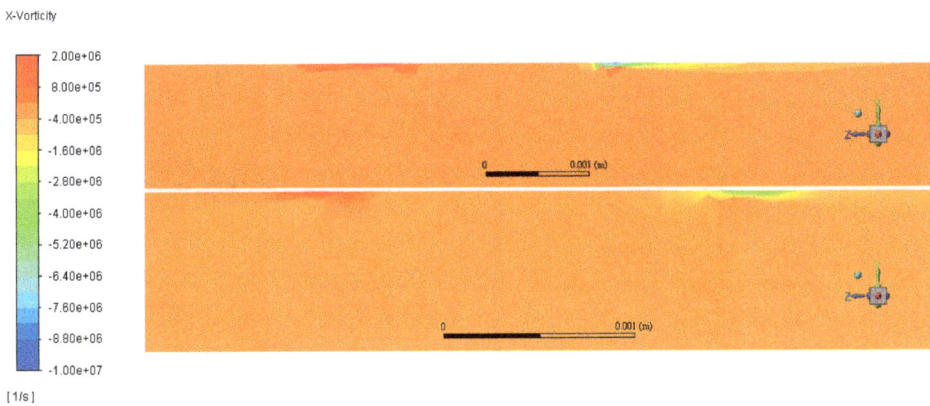

Figure 6:   x-vorticity contour field around the leak, case 3, time = 10 s (upper: air; lower: methane).

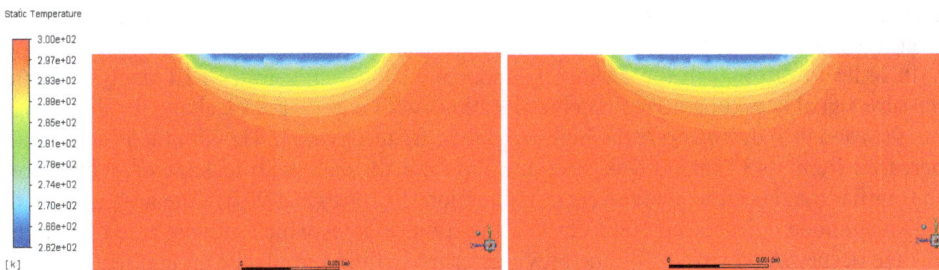

Figure 7:   Temperature contours around the leak at the yz plane, time = 10 s (left: air; right: methane).

temperature drops around the leak orifice. Near the leak vicinity a strong vortex exists and results complex flow field around the leak. Due to this complex flow field around the leak small eddies dissipate, and the energy turns from kinetic energy to thermal energy, cause local warming effect near leakage region.

Fig. 8 shows the static pressures distribution along the various line locations along the pipeline. Fig. 8(a) illustrates the sudden pressure drop at the leak direction along the line which is 1 mm below the leak point and midplane, respectively. The pressure drop is higher closer to the leak point and there is a slight pressure increases after the leak area. Fig. 8(b) indicates the difference between the case with and without leak for case 2. The reason of the kink at the pressure data, there is a circulation created due to leak region. After the leak region, velocity of the fluid decreases and local pressure is increased, suddenly.

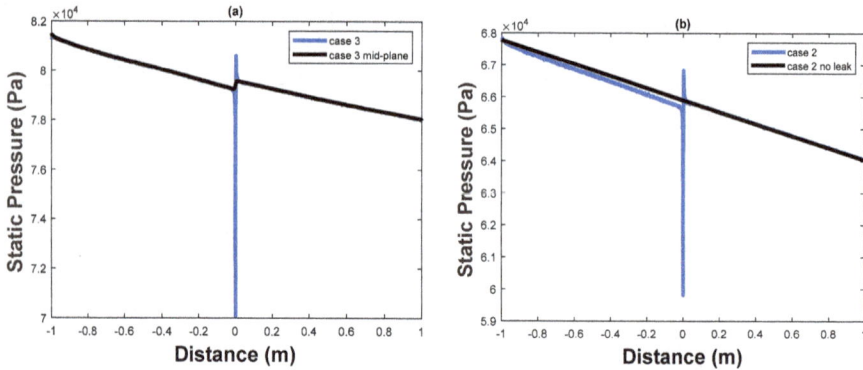

Figure 8:   Pressure variation along the straight line, time = 10 s. (a) 1 mm and 2 mm below the leak for case 3; and (b) 1 mm below the leak and no leak conditions for case 2.

3.6 FFT analysis

The Fast Fourier Transform (FFT) and power spectral density (PSD) analysis are performing on the pressure signal to understand the transient turbulent behavior of the air flow field. In this study, Delayed Detached Eddy Simulation (DES) is conducted due to fact it is less expensive than the Large Eddy Simulation (LES), but still predict flow and turbulence distribution well around the leak location. Pressure fluctuations are monitored at several receiver locations.

In order to distinguish the leakage phenomenon, Fig. 9 illustrates FFT analysis of the pressure signal at prementioned receiver locations with and without leakage for the case 3. Fig. 9(a) and 9(b) demonstrate the high amplitude frequency at 31 Hz with the PSD value is increased from 1,437 for no leak case to 12,610 for the case with leakage on y direction. The similar trend occurs on receivers at z-direction (Fig. 9(c) and 9(d)). The high amount of PSD magnitude is increased due to fact of the sonic sources which are generated when the leakage occurs. As discussed on the previous chapter, fluid turbulence around the leak area where gas jets out from the hole causes the sonic waves. Fig. 10(a) and 10(b) indicates the PSD value is increased from 1,283 for no leak case to 2,460 for the case with leakage on y direction. The similar trend can be monitored in Fig. 10(c) and 10(d). Moreover, the peak value, 31 Hz, shows the main energy of the acoustic pressure at low frequency. Due to fact that low frequency can transfer long distance, it is easier to detect this signal within the short time.

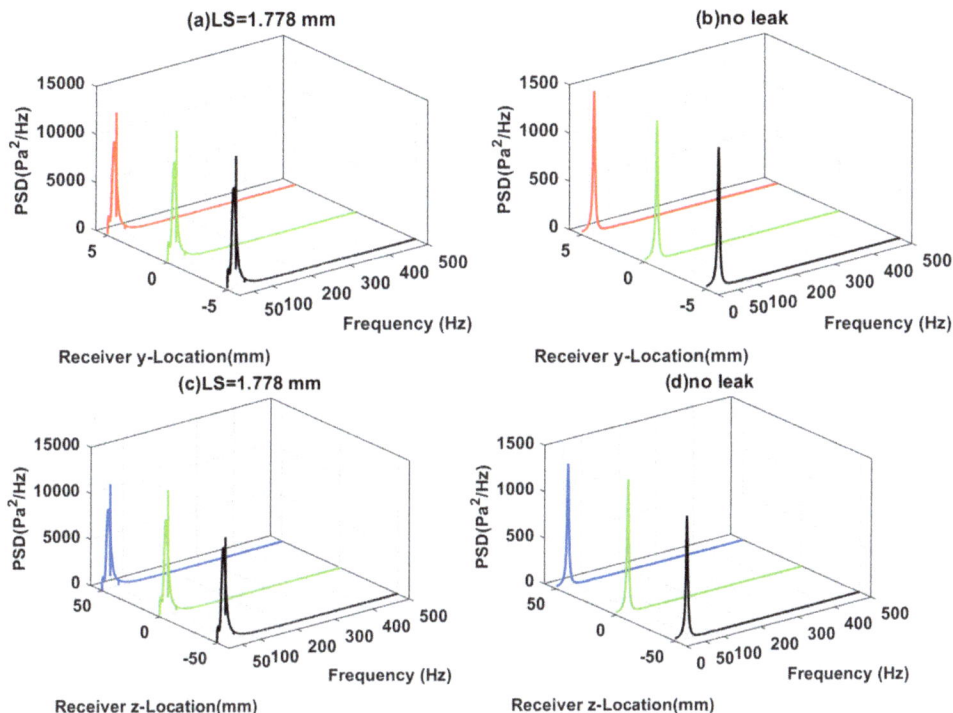

Figure 9:   FFT of pressure signal at various receiver locations for case of leak (left), and with no leak (right)-case 3.

4 CONCLUSION

The main objective of this study is to determine the observable trends of pinhole leaks in gas pipelines. Experimental setup is established to implement of pressure point analysis PPA for leak detection. Also, three-dimensional, transient CFD turbulent flow simulations are performed to investigate the unsteady behaviors of the fluid domain in the near region of the leak. We can summarize the conclusions as follows:

1. Leak mass flow rate is calculated from the mass balance method at the steady conditions by experimental setup. For low gas pressure application, it is an accurate method to use.
2. Pressure point analysis (PPA) method is used to monitor dynamic and differential pressure at before and after the leak point to compare the previous pressure measurements. This method is cost effective and easy to maintain.
3. Gas leak causes high peak of the acoustic pressure signal for the range of 31 Hz frequency. Acoustic energy of pressures causes high peak on the low frequency, due to fact that low frequency can transfer long distance, it is easier to detect this signal within the short time.
4. Acoustic energy analysis on different positions shows there is a correlation between the signal's magnitude and leak location. More simulations are needed to develop a precise model to foresee effects and outcome of future accidents caused by the hazardous material from the pipeline leak.

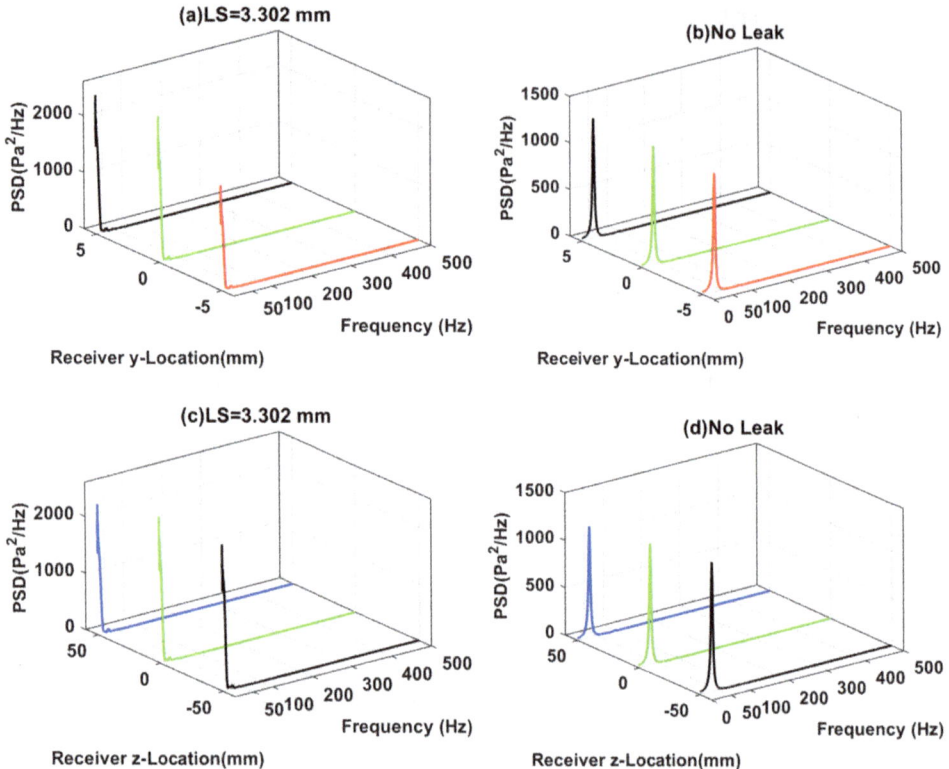

Figure 10: FFT of pressure signal at various receiver locations for case of leak (left), and with no leak (right)-case 2.

All aforementioned methods are used for detecting leaks in gas pipeline systems in offshore application. Future studies will focus on the accurately localization of the leak by using the above methods.

ACKNOWLEDGEMENTS

The HPC (High Performance Computing) resources and services used in this work were provided by the Research Computing group in Texas A&M University at Qatar. Research Computing is funded by the Qatar Foundation for Education, Science and Community Development. This publication was made possible by the Responsive Research Seed Grants (RRSG) funding provided by Texas A&M University at Qatar.

REFERENCES

[1] Behari, N., Sheriff, M.Z., Rahman, M.A., Nounou, M., Hassan, I. & Nounou, H., Chronic leak detection for single and multiphase flow: A critical review on onshore and offshore subsea and Arctic conditions. *J. Nat. Gas Sci. Eng.*, **81**, Elsevier B.V., 01 Sep. 2020. DOI: 10.1016/j.jngse.2020.103460.

[2] Joliffe, I.T., *Principal Component Analysis*, 2nd edn, Springer-Verlag: New York, 2002.

[3]   Ben-Mansour, R., Habib, M.A., Khalifa, A., Youcef-Toumi, K. & Chatzigeorgiou, D., Computational fluid dynamic simulation of small leaks in water pipelines for direct leak pressure transduction. *Comput. Fluids*, **57**, pp. 110–123, 2012. DOI: 10.1016/j.compfluid.2011.12.016.

[4]   Ben-Mansour, R., Suare, K.A. & Youcef-Toumi, K., Determination of important flow characteristics for leak detection in water pipelines-networks. *9th International Conference on Heat Transfer, Fluid Mechanics and Thermodynamics*, Malta, 2012.

[5]   Liu, C., Li, Y., Meng, L., Wang, W. & Zhang, F., Study on leak-acoustics generation mechanism for natural gas pipelines. *J. Loss Prev. Process Ind.*, **32**, pp. 174–181, 2014. DOI: 10.1016/j.jlp.2014.08.010.

[6]   Liu, C., Wang, Y., Li, Y. & Xu, M., Experimental study on new leak location methods for natural gas pipelines based on dynamic pressure waves. *J. Nat. Gas Sci. Eng.*, **54**(Mar.), pp. 83–91, 2018. DOI: 10.1016/j.jngse.2018.03.023.

[7]   Xiong, X., Rahman, M.A. & Zhang, Y., RANS based computational fluid dynamics simulation of fully developed turbulent Newtonian flow in concentric annuli. *J. Fluids Eng. Trans. ASME*, **138**(9), pp. 1–9, 2016. DOI: 10.1115/1.4033314.

[8]   Amin, A., Imtiaz, S., Rahman, A. & Khan, F., Nonlinear model predictive control of a Hammerstein Weiner model based experimental managed pressure drilling setup. *ISA Trans.*, **88**, pp. 225–232, 2019. DOI: 10.1016/j.isatra.2018.12.008.

[9]   Rahman, M.A., Mustafiz, S., Biazar, J., Koksal, M. & Islam, M.R., Investigation of a novel perforation technique in petroleum wells-perforation by drilling. *J. Franklin Inst.*, **344**(5), pp. 777–789, 2007. DOI: 10.1016/j.jfranklin.2006.05.001.

[10]   Rahman, M.A., Heidrick, T. & Fleck, B.A., Characterizing the two-phase, air/liquid spray profile using a phase-doppler-particle-analyzer. *IOP J. Phys. Conf. Ser.*, **147**(Jan.), pp. 1–15, 2009.

[11]   Rahman, M.A., Heidrick, T. & Fleck, B.A., A critical review of two-phase gas-liquid industrial spray systems. *Int. Rev. Mech. Eng.*, **3**(1), pp. 110–125, 2009.

[12]   Ahammad, M.J., Rahman, M.A., Zheng, L., Alam, J.M. & Butt, S.D., Numerical investigation of two-phase fluid flow in a perforation tunnel. *J. Nat. Gas Sci. Eng.*, **55**(Nov.), 2017, pp. 606–611, 2018. DOI: 10.1016/j.jngse.2017.10.016.

[13]   Sleiti, A.K., Takalkar, G., El-Naas, M.H., Hasan, A.R. & Rahman, M.A., Early gas kick detection in vertical wells via transient multiphase flow modelling: A review. *J. Nat. Gas Sci. Eng.*, **80**(May), p. 103391, 2020. DOI: 10.1016/j.jngse.2020.103391.

[14]   Manikonda, K., Hasan, A.R., Kaldirim, O., Schubert, J.J. & Rahman, M.A., Understanding gas kick behavior in water and oil-based drilling fluids. *SPE Kuwait Oil Gas Show Conference, 2019, KOGS 2019*, 2019. DOI: 10.2118/198069-ms.

[15]   Herath, D., Khan, F., Rathnayaka, S. & Rahman, M.A., Probabilistic estimation of hydrate formation. *J. Pet. Sci. Eng.*, **135**, pp. 32–38, 2015. DOI: 10.1016/j.petrol.2015.08.007.

[16]   Qureshi, M.F., Ali, M., Rahman, M.A., Hassan, I., Rasul, G. & Hassan, R., Experimental investigation of multi-phase flow in an annulus using electric resistance tomography. *SPE Kuwait Oil Gas Show Conference, 2019, KOGS 2019*, (Mar.), 2018, pp. 1947–1956, 2019. DOI: 10.2118/198011-ms.

[17]   Zahid, A.A., ur Rehman, S.R., Rushd, S., Hasan, A. & Rahman, M.A., Experimental investigation of multiphase flow behavior in drilling annuli using high speed visualization technique. *Front. Energy*, **14**(3), pp. 635–643, 2020. DOI: 10.1007/s11708-018-0582-y.

[18] Jujuly, M., Thodi, P., Rahman, A. & Khan, F., Computational fluid dynamics modeling of subsea pipeline leaks in Arctic conditions. *Arctic Technology Conference Offshore Technology Conference*, St. John's, Newfoundland and Labrador, Canada, p. 19, 2016. DOI: 10.4043/27417-MS.

[19] Menter, F.R., Kuntz, M. & Langtry, R., Ten years of industrial experience with the SST turbulence model turbulence heat and mass transfer. *Cfd.Spbstu.Ru*, **4**(Jul.), 2014, pp. 625–632, 2003.

[20] Gritskevich, M.S., Garbaruk, A.V., Schütze, J. & Menter, F.R., Development of DDES and IDDES formulations for the k-ω shear stress transport model. *Flow. Turbul. Combust.*, **88**(3), pp. 431–449, 2012. DOI: 10.1007/s10494-011-9378-4.

[21] Spalart, P.R., Deck, S., Shur, M.L., Squires, K.D., Strelets, M.K. & Travin, A., A new version of detached-eddy simulation, resistant to ambiguous grid densities. *Theor. Comput. Fluid Dyn.*, **20**(3), pp. 181–195, 2006. DOI: 10.1007/s00162-006-0015-0.

[22] Versteeg, H.K. & Malalasekera, W. (eds), *An Introduction to Computational Fluid Dynamics: The Finite Volume Method*, 2nd edn, Pearson Education Ltd., 2007.

COMPUTATIONAL AND EXPERIMENTAL STUDY ON RING TONE

KAZUO MATSUURA[1], KOH MUKAI[1] & MIKAEL ANDERSEN LANGTHJEM[2]
[1]Graduate School of Science and Engineering, Ehime University, Japan
[2]Department of Engineering, Aarhus University, Denmark

ABSTRACT

Sound produced when jets issued from a circular nozzle collide with a downstream ring coaxial with the jet is called ring tone. The ring tone is investigated through experiments and direct sound computations. The inner diameters of the circular nozzle exit and the ring are both 30 mm. In the experiments, the frequency spectra of the ring tone are measured for various impingement lengths of 20–40 mm and jet velocities of 5–15 m/s. The tone is also compared with the hole tone, in which the downstream ring is replaced with a plate with a hole. As a general trend, the peak sound intensity shifts to higher frequency as the jet speed is increased in both tones. Multiple series of peaks are observed for each impingement length. While the peaks generally become higher with increasing jet speed, the amplification is not necessarily monotonic. When the ring tone is compared with the hole tone, the peaks are much lower and the frequency distribution is broader. Regarding the variation of the dominant ring-tone and hole-tone peak frequencies with jet speed, differences between the ring tone and hole tone appear at the locations of mode jumps. In the computations, the onset of self-sustained feedback oscillations in the ring tone is clarified from the view point of the throttling mechanism originally proposed for the hole tone, i.e., the coupling between the mass flow through the ring, vortex impingement and global pressure fluctuation.
Keywords: ring tone, hole tone, aeroacoustics, direct sound computation, wind tunnel.

1 INTRODUCTION

Ring tone is a kind of feedback sound emitted from a system in which pressure waves generated when vortices from a nozzle collide with a downstream obstacle, propagate upstream, and regulate the timing of further vortex ejection from the nozzle [1]–[4]. We encounter feedback tones in many practical situations such as solid propellant rocket motors, automobile intake and exhaust systems, ventilation systems, gas distribution systems, and whistling kettles.

Ring tone is a dipole sound with an efficiency scaling of the third power of the jet Mach number; the dipole strength is equal to the unsteady force on the ring [4]. These dipoles radiate in the direction of the jet axis [3]. Chanaud and Powell [5] investigated ring tone with a torus with an inner diameter of 5 mm formed from copper wire with a diameter of 1.5 mm. Data were taken by varying the edge distance at constant Reynolds number and by varying the Reynolds number at constant edge distance. They found that at large spacing ratios, the hole- and ring-tone minimum-edge-distance contours have similar shapes and are distinct from an edge-tone contour. At small spacing ratios, the minimum-edge-distance contour for the ring-tone system deviates markedly from other contours, there being folds in the contours. The folds are seen to be related to jumps at higher Reynolds numbers. Obata et al. [6] experimentally investigated the ring tone. They found that multiple frequency components have substantial amplitudes compared to the fundamental frequency component, and showed that the lowest frequency component can interact with the fundamental component either to reinforce itself or to produce an additional frequency component.

Although edge and hole tones have been investigated relatively in detail [3], [7]–[10] among feedback sounds, the details of ring tones, such as flow fields and the difference

WIT Transactions on Engineering Sciences, Vol 132, © 2021 WIT Press
www.witpress.com, ISSN 1743-3533 (on-line)
doi:10.2495/MPF210121

between the ring and hole tones, are unknown except for the description mentioned above. Therefore, in this study, a ring tone system is developed, the frequency spectra of the ring tone are measured for various impingement lengths and jet velocities, and the sound generation mechanism is investigated through a direct sound computation. In the measurements, the ring tone is compared with the hole tone, which is generated when a metal plate with a hole with the same diameter as the nozzle outlet is installed instead of the ring with the ring tone. To our best knowledge, this study is the first case of applying a direct computation to the ring tone.

In Section 2, the experimental methods are described including the geometries of the ring and hole, the wind tunnel, flow conditions, and the measurement system. Section 3 describes the numerical methods, computational conditions, grids, and initial and boundary conditions. In Section 4, experimental results are shown such as frequency spectra and the variation of the dominant peak frequency for various jet speeds and impingement lengths. The experimental results for the ring tone are compared with those for the hole tone. Section 5 discusses the relationship between vortex impingement on the ring, mass flow variation through the ring, and global pressure variation in the flow field between the nozzle exit and the ring.

2 EXPERIMENTAL METHODS

Experiments were conducted to gain an insight into the sound properties of the ring tone. Fig. 1 shows the experimental system with a ring placed downstream of the nozzle. A jet issued from the circular nozzle of a wind tunnel, and the ring was coaxial to the nozzle. The inner diameters of the nozzle exit d_0 and the ring were both 30 mm. The outer diameter of the ring was 40 mm. The distance between the nozzle exit and the ring, the impingement length L_{im} was varied from 20 to 40 mm. Sound pressure was measured at $r = 120$ mm for $z = L_{im}/2$ with a condenser microphone. Data were passed to an FFT analyser. The ring was acoustically compact. The speed of the circular jet was varied from 5 to 15 m/s. At 20°C, these velocities correspond to Reynolds numbers ($Re = u_0 d_0/\nu$) of 9.96×10^3–2.98×10^4. Fig. 2 shows the overview and schematic of the wind tunnel. The circular jets were generated by a smooth circular contraction of airflow from the almost axisymmetric wind tunnel.

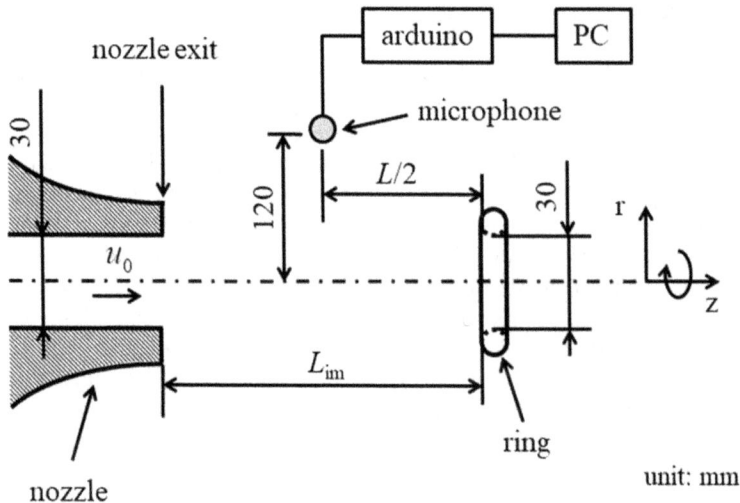

Figure 1: Experimental system.

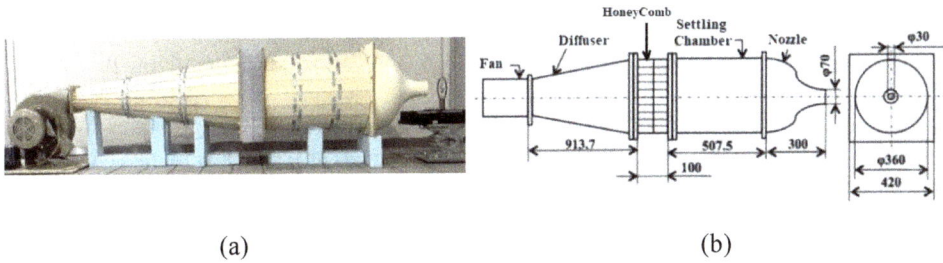

(a)   (b)

Figure 2: Wind tunnel. (a) Photo; and (b) Schematic; unit: mm.

In addition to the experiments on the ring tone, experiments on the hole tone were conducted to compare the sound properties of the two tones because the hole tone is similar and more details are known about it. Fig. 3 shows the fabricated ring and hole plate and their schematics. To fix the ring in the air, the ring was pulled by six piano wires each with a diameter of 0.25 mm.

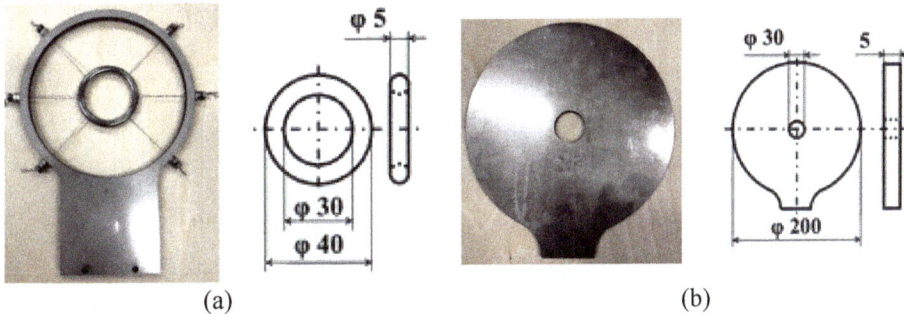

(a)   (b)

Figure 3:   Ring and hole plate; unit: mm. The inner and outer diameters of the ring are 30 mm and 40 mm, respectively. The diameter of the hole is 30 mm, and the thickness is 5 mm. (a) Fabricated ring (left: photo; right: schematic); and (b) Fabricated hole plate (left: photo; right: schematic).

3 NUMERICAL METHODS

A direct sound computation was conducted to clarify the onset of the self-sustained oscillation of the ring tone. Only near-field behaviours were considered because the interaction of a flow with the ring and the associated feedback propagation of pressure to the nozzle exit is the central concern of the computation.

The governing equations are the unsteady three-dimensional compressible Navier–Stokes equations. The ideal gas law closes the system of the equations. The equations were solved with the finite-difference method. Spatial derivatives that appear in metrics, convective and viscous terms were evaluated with the 6th-order tridiagonal compact scheme [11]. Near the boundaries, the 4th-order one-sided classical Padé scheme was used at one point inside the boundaries. Time-accurate solutions to the governing equations were obtained using the 3rd-order Runge–Kutta scheme. In addition to the spatial discretisation and time integration, 10th-order implicit filtering [12] was introduced to suppress numerical instabilities that arise from the central differencing in the compact scheme. The parameter on the left-hand side, which is associated with filtering strength, was set to be 0.492. An implicit 4th-order filter was used near the boundaries. This method has been well validated for predicting the original hole tone [9], [10].

Fig. 4 shows the structured mesh used for the computation. Every five grid lines are drawn for clarity. There were a total of approximately 5.2 M grid points. The overall mesh consisted of four zones: the inlet nozzle, the exterior domain of the nozzle, the ring zone, and the background space. Each zone is shown in a different colour. The ring in the computational domains was represented with the overset method. The outer boundaries of the ring zone were interpolated from the background space at each time step. The boundaries of the blank region of the background space were interpolated from the ring zone. Trilinear interpolation was used in this study. The z-axis was defined in the streamwise (axial) direction. Near the walls, the first mesh widths in the normal directions were $8.33 \times 10^{-4} d_0$. The streamwise widths of the background mesh around the ring were approximately $3.28 \times 10^{-2} d_0$. The far boundaries in the z and r directions were located at $70.36 d_0$ and $70.86 d_0$, respectively. The inlet boundary of the inlet nozzle was located at $z = -0.6 d_0$.

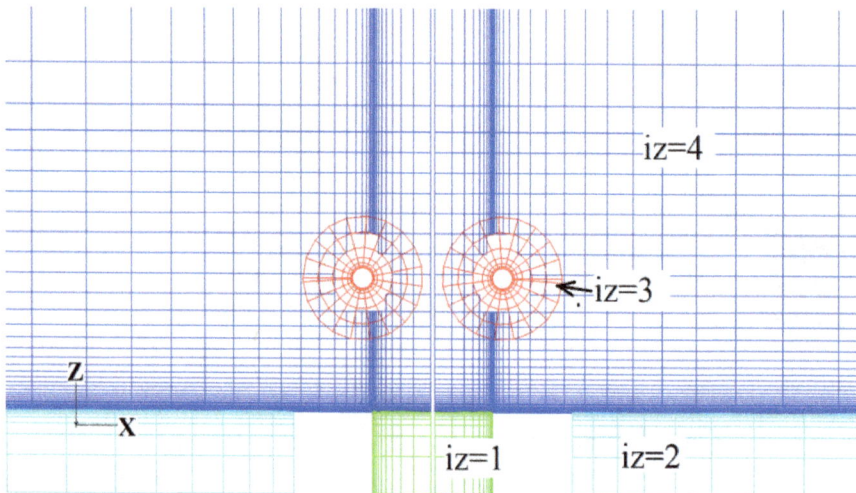

Figure 4:   Computational mesh for the ring tone (axisymmetric). Zones iz = 1,...,4 are the nozzle inlet, the exterior domain of the nozzle, the ring zone, and the background space, respectively. Every five grid lines are drawn for clarity.

4 EXPERIMENTAL RESULTS

The frequency spectra of the ring tone were measured for various impingement lengths and jet velocities [13]. Similar measurements were conducted also for the hole tone to clarify the differences between the ring and hole tones.

Fig. 5 shows the results. As a general trend, the peak sound intensity shifts to higher frequency with increasing jet speed in both tones. Multiple series of peaks are observed for L_{im} in both tones. While the peak sound intensity generally becomes stronger with increasing jet speed, the amplification is not monotonic. When the ring tone is compared with the hole tone, the peak sound intensities are much lower and the frequency distribution is broader. The lower intensity of the ring tone is consistent with [14].

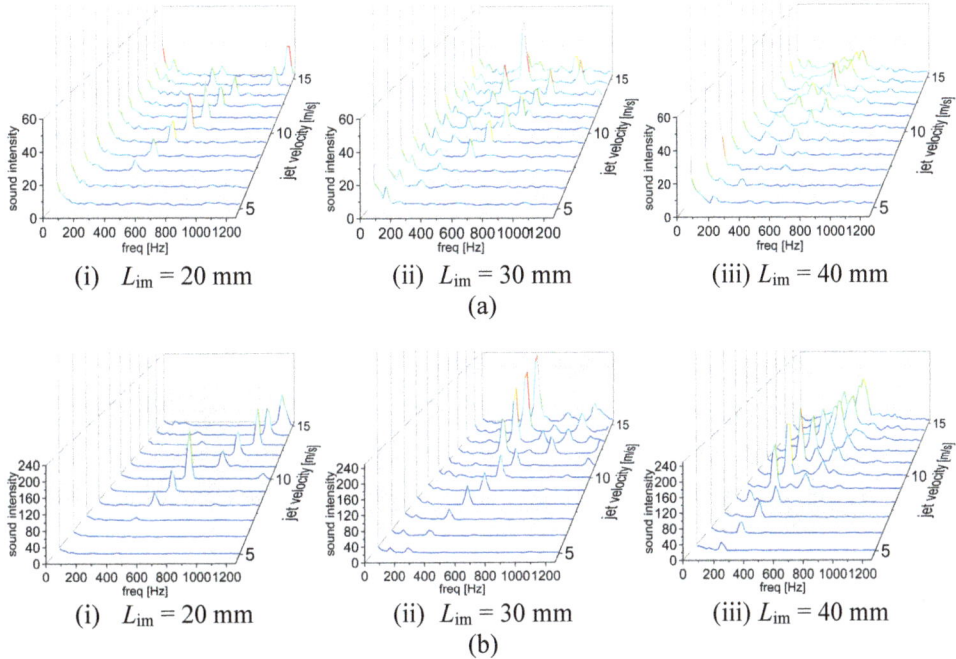

(i) $L_{im} = 20$ mm   (ii) $L_{im} = 30$ mm   (iii) $L_{im} = 40$ mm

(a)

(i) $L_{im} = 20$ mm   (ii) $L_{im} = 30$ mm   (iii) $L_{im} = 40$ mm

(b)

Figure 5:   Frequency spectra for various impingement lengths and jet speeds. (a) Ring tone; and (b) Hole tone.

Fig. 6 shows the variation of the dominant ring-tone and hole-tone peak frequencies with jet speed for $L_{im} = 20$, 30 and 40 mm. In the figure, correlations based on Rossiter's equations $n/f = L_{im}/u_c + L_{im}/c_0$ [15] are also plotted. Here, n, f, L_{im}, and c_0 are the stage index, frequency, impingement length, and speed of sound, respectively; $u_c = 0.6u_0$ is the convection velocity. As with the computational results in Fig. 5, the peak frequencies generally increase linearly. Differences between the ring tone and the hole tone appear at the locations of mode jumps, i.e., jumps in the stage index. Both integer stage indices (n) and fractional indices appear.

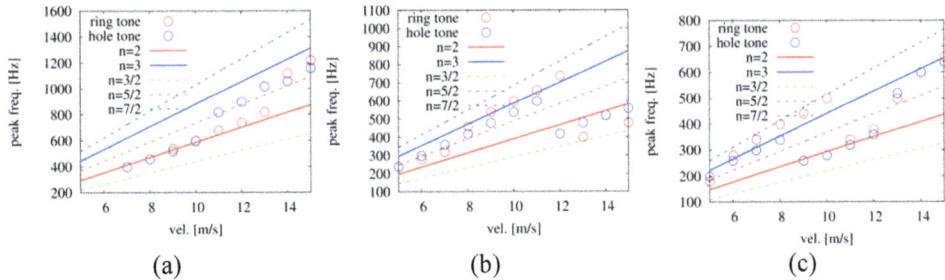

(a)   (b)   (c)

Figure 6:   Variation of the dominant peak frequency with jet speed for various impingement lengths for the ring tone and hole tone. (a) $L_{im} = 20$ mm; (b) $L_{im} = 30$ mm; and (c) $L_{im} = 40$ mm.

Fig. 7 shows vortex structures and the divergence of velocity at different times. The vortical structures are visualised with the iso-surfaces of the second invariance of the velocity gradient tensor for $Q = 1 \times 10^{-5}$. t denotes the elapsed time from the initial condition, and t_d is the time corresponding to 2,500 computational time steps. $t = 0$ corresponds to the initial condition with quiescent air in the flow field except for the nozzle inlet, where a Pohlhausen-type laminar velocity profile is imposed. The boundary layer thickness is assumed to be 1 mm. Velocities and lengths are non-dimensionalised with reference quantities v_{ref} and l_{ref}, respectively. In this study, $v_{ref} = 340$ m/s and $l_{ref} = 1$ mm. Vortices shed from the nozzle exit collide with the ring. When the vortices pass through the ring, new vortices are generated.

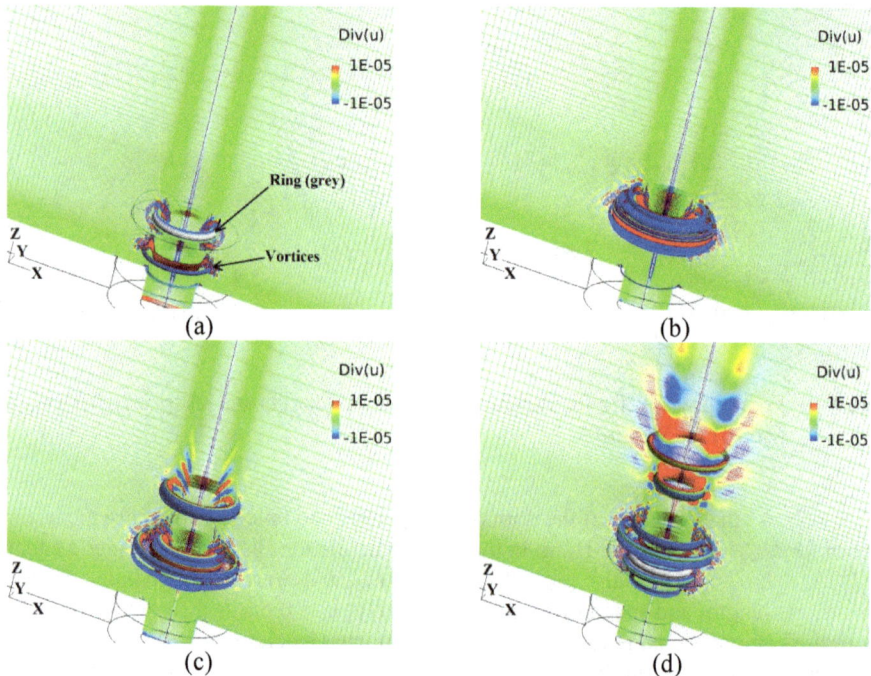

Figure 7: Vortical structures and divergence of velocity. The vortical structures are visualised with iso-surfaces of the second invariance of the velocity gradient tensor for $Q = 1 \times 10^{-5}$. (a) $t/t_d = 34$; (b) $t/t_d = 80$; (c) $t/t_d = 126$; and (d) $t/t_d = 174$.

To show the interaction of vortices with the ring more clearly, Fig. 8 shows the distribution of ω_y in the $y = 0$ plane. Here, the y direction is normal to the sheet. At $t/t_d = 34$, a vortex ring is convected along the circular jet and located between the nozzle exit and the ring. At $t/t_d = 80$, the initial vortex ring collides with the ring, and secondary vortices, i.e., the wakes of the ring, are generated. After the vortices pass through the ring, some vortices are convected rapidly, and other vortices stay around the ring for a while and are convected afterward.

The transient process from the initial state to the oscillatory states was analysed to investigate the onset of the feedback oscillation. As clarified in [9], the hole tone is governed

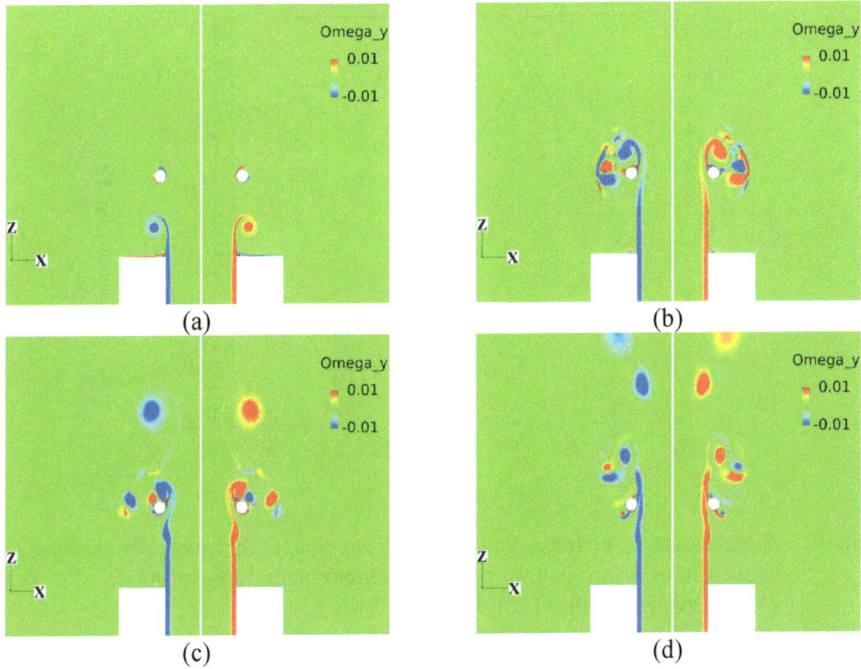

Figure 8:   Distribution of ω_y on $y = 0$ plane. (a) $t/t_d = 34$; (b) $t/t_d = 80$; (c) $t/t_d = 126$; and (d) $t/t_d = 174$.

by the axisymmetric throttling mechanism, which links mass flow through the hole, vortex impingement and global pressure fluctuation. The ring tone was analysed from the same viewpoint.

Fig. 9 shows the time variation of the mass flow through the ring and gauge pressure sampled at P_1 [$(r,z) = (15, 0.126)$], P_2 [$(7.46, 0.126)$], and P_3 [$(22.5, 0.126)$]. The axial position of the mass flow evaluation is the centre of the ring. Initially, the mass flow increases monotonically until around $t/t_d = 72$. Oscillations in the mass flow appear afterwards. Peaks are observed at $t/t_d = 72$, 119, and 145, and valleys are observed at $t/t_d = 95$, 134, and 155. Except for the initial peak around $t/t_d = 56$, the peaks and valleys of the gauge pressure approximately coincide with those of the mass flow variation, which means that mass flow variation induces the pressure oscillations. The time for sound propagation over $L_{im} = 30$ mm is $\Delta t/t_d = 0.625$. Therefore, the time lag of the pressure fluctuation between the ring and the nozzle exit is negligible.

To clarify velocity fields causing the mass flow variation through the ring, Fig. 10 shows the distributions at the peaks and valleys of circumferentially-averaged z-directional and r-directional velocities along the radial position from the rotational axis. The velocity vector v at a point is decomposed into the radial (r), axial (z), and circumferential (θ) components using cylindrical coordinates: $v = v_r e_r + v_\theta e_\theta + v_z e_z$, where e_r, e_θ, and e_z are the unit normal basis vectors. The downstream direction of the rotational axis is taken as the positive z direction, and e_z and e_r are taken in the positive z direction and outward directions, respectively. The circumferential direction is then defined according to a right-handed system.

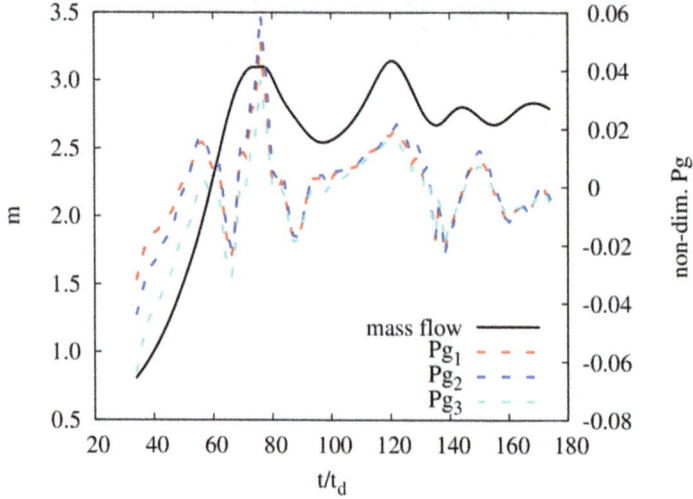

Figure 9:   Time variation of mass flow through the ring m and gauge pressure P_g. The subscripts of the gauge pressure denote sampling locations. P_1: (r,z) = $(15, 0.126)$; P_2: $(7.46, 0.126)$; P_3: $(22.5, 0.126)$.

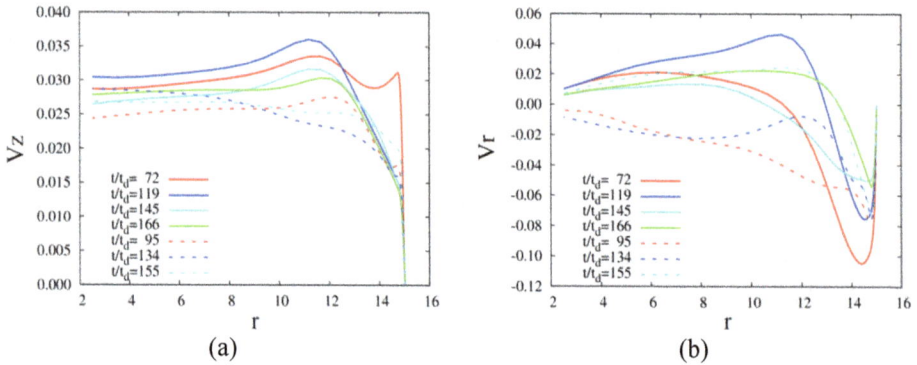

Figure 10: Circumferentially-averaged velocities in the axial (z) and radial (r) directions. r is the radial distance from the rotational axis. (a) z-directional velocity; and (b) r-directional velocity.

At the peaks, v_z and v_r are generally higher than at the valleys. The rapid drops in v_z and v_r near $r = d_0/2$ mean the existence of a wall. Positive v_r means that flow vectors point outward when the flow passes through the ring, and negative v_r means that flow vectors point inward. Thus, the periodic variation of flow vectors causes mass flow variation through the ring. Fig. 11 shows the distribution of ω_y when the mass flows are the peaks and valleys. As observed in Fig. 11(a), the bulged regions denoted by the black arrows leave the ring when the mass flows become peaks. When the mass flows become valleys, the bulged regions collide with the ring as found in Fig. 11(b).

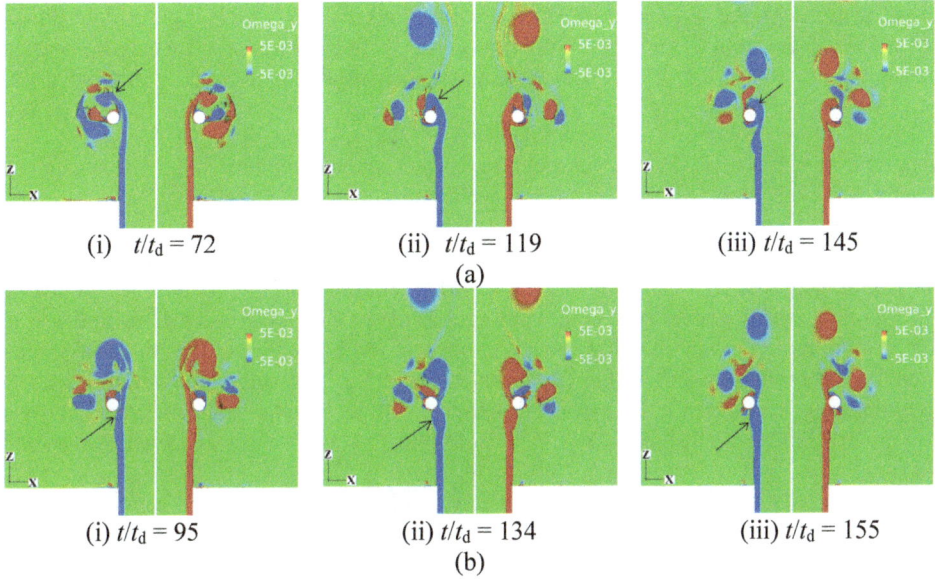

(i) $t/t_d = 72$ (ii) $t/t_d = 119$ (iii) $t/t_d = 145$
(a)

(i) $t/t_d = 95$ (ii) $t/t_d = 134$ (iii) $t/t_d = 155$
(b)

Figure 11: Distribution of ω_y when mass flows through the ring are peaks and valleys. (a) At peaks; and (b) At valleys.

Fig. 12 shows the distributions of P_g/p_∞ when the mass flows through the rings are peaks and valleys. Although the pressure becomes high in the regions around the jet connecting the nozzle exit and the ring at the peaks, it becomes low at the valleys. Jet fluid is pushed and

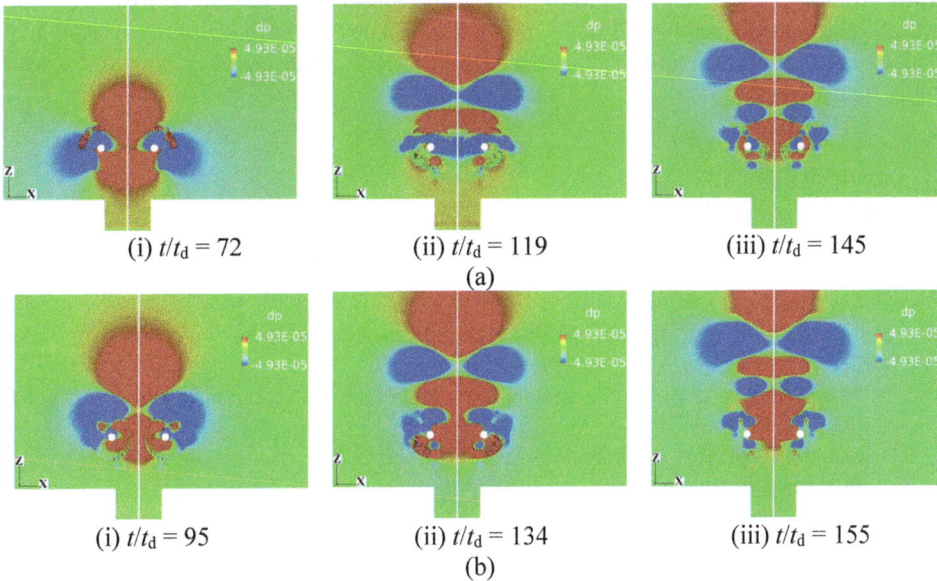

(i) $t/t_d = 72$ (ii) $t/t_d = 119$ (iii) $t/t_d = 145$
(a)

(i) $t/t_d = 95$ (ii) $t/t_d = 134$ (iii) $t/t_d = 155$
(b)

Figure 12: Distribution of P_g/p_∞ when mass flows through the ring are peaks and valleys. (a) At peaks; and (b) At valleys.

mass flow through the ring becomes large when the ambient pressure is high, and jet fluid is pulled back (decelerated) and mass flow through the ring becomes small when the ambient pressure is low. Thus, this mechanism generates a self-sustained feedback oscillation.

5 CONCLUSIONS

Ring tone was investigated via experiments and direct sound computation. In the experiments, the frequency spectra of the ring tone were measured for various jet velocities and impingement lengths. The ring tone was also compared with the hole tone.

The peak sound intensity generally shifts to higher frequency as the jet speed increases in both tones. Multiple series of peaks are observed at a particular L_{im} in both tones. While the peak sound intensity generally becomes stronger as the jet speed increases, the amplification is not monotonic. When the ring tone is compared with the hole tone, the peak sound intensities are much lower and the frequency distribution is broader. The variations of the dominant ring-tone and hole-tone peak frequencies with jet speed reveal differences between the ring and hole tones at the mode jumps. Both integer stage indices and fractional indices appear.

The computation investigated the onset of the self-sustained feedback oscillation in the ring tone from the view point of the throttling mechanism. The mass flows through the ring are regulated by the variation in flow velocity and angle because of the interaction of a vortex ring around a sinuously deformed jet with the obstacle ring. This includes flow separation and the generation of secondary vortices. The mass flow becomes large when the ambient pressure in the region extending from the nozzle exit and the ring is high, and becomes small when the ambient pressure is low. Thus, the pressure fluctuation perturbs the nozzle exit where discrete vortices are generated.

ACKNOWLEDGEMENTS

Computations were partly conducted using a supercomputer system at the Japan Aerospace Exploration Agency (JAXA-JSS3). The authors thank Shin Nippon Feather Core Co., Ltd. for providing us honeycomb meshes for the wind tunnel developed in this study. The authors thank Mr. Mark Kurban from Edanz Group (https://www.jp.edanz.com/ac) for editing a draft of this manuscript.

REFERENCES

[1] Rayleigh, L., *Theory of Sound*, vol. 2, Dover Publication, pp. 410–412, 1945.
[2] Rockwell, D. & Naudascher, E., Self-sustained oscillations of impinging free shear layers. *Ann. Rev. Fluid Mech.*, **11**, pp. 67–94, 1979.
[3] Blake, W.K., *Mechanics of Flow-Induced Sound and Vibration*, vol. 1, Academic Press, Inc., p. 151, 1986.
[4] Howe, M.S., *Acoustics of Fluid-Structure Interactions*, Cambridge University Press, 1998.
[5] Chanaud, R.C. & Powell, A., Some experiments concerning the hole and ring tone. *J. Acoust. Soc. Am.*, **37**(5), pp. 902–911, 1965.
[6] Obata, T., Kurasawa, H. & Haneda, Y., Self-excited oscillation in an axisymmetric jet with a coaxial ring (in Japanese). *Trans. Jpn. Soc. Mech. Eng. B.*, **61**(583), No. 94–1045, pp. 106–112, 1995.
[7] Langthjem, M.A. & Nakano, M., A numerical simulation of the hole-tone feedback cycle based on an axisymmetric discrete vortex method and Curle's equation. *J. Sound Vib.*, **288**(1–2), pp. 133–176, 2005.

[8]   Langthjem, M.A. & Nakano, M., A three-dimensional study of the hole-tone feedback problem. *RIMS Kôkyûroku*, **1697**, pp. 80–94, 2010.

[9]   Matsuura, K. & Nakano, M., Direct computation of a hole-tone feedback system at very low Mach numbers. *J. Fluid Sci. Tech.*, **6**(4), pp. 548–561, 2011.

[10] Matsuura, K. & Nakano, M., A throttling mechanism sustaining a hole tone feedback system at very low Mach numbers. *J. Fluid Mech.*, **710**, pp. 569–605, 2012.

[11] Lele, S.K., Compact finite difference schemes with spectral-like resolution. *J. Comput. Phys.*, **103**(1), pp. 16–42, 1992.

[12] Gaitonde, D.V. & Visbal, M.R., Padé-type higher-order boundary filters for the Navier–Stokes equations. *AIAA J.*, **38**(11), pp. 2103–2112, 2000.

[13] Mukai, K. & Matsuura, K., Experiments and computations on ring tone (in Japanese). *Proceedings of the JSME, Chushikoku, 59th General Assembly Meeting*, pp. 1–2, 2021.

[14] Powell, A., Some aspects of aeroacoustics: From Rayleigh until today. *J. Vib. Acoust.*, **112**, pp. 145–159, 1990.

[15] Rossiter, J.E., The effect of cavities on the buffeting of aircraft. *RAE Tech. Memorandum*, **754**, 1962.

ULTRASONIC MEASUREMENT OF MULTI-DIMENSIONAL VELOCITY VECTOR PROFILE USING ARRAY TRANSDUCER

NARUKI SHOJI[1], HIDEHARU TAKAHASHI[2] & HIROSHIGE KIKURA[2]
[1]Department of Mechanical Engineering, Tokyo Institute of Technology, Japan
[2]Laboratory for Zero-Carbon Energy, Tokyo Institute of Technology, Japan

ABSTRACT

In March 2011, the severe accident of the Fukushima Dai-ichi (1F) nuclear power plant was happened by the earth quack and massive tsunami in Tohoku, Japan. And then, fuel debris was generated within the primary containment vessels (PCVs) of units 1, 2, and 3, respectively. Recently, the decommissioning of 1F is underway to remove the fuel debris, and the inside inspection with robots was conducted so far. Optical techniques have been applied for inspecting the PCVs, but information of the contaminated water leakage has been not unveiled due to non-clear water causing poor visibility of the camera. Therefore, non-optical techniques are required to unveil the leakage location, and we focused on the ultrasonic measurement technique. Ultrasonic measurement can be applied to opaque liquid, high-radioactive, and dark environments. In this study, we have developed the ultrasonic velocity profiler (UVP) system for the investigation of leaking locations. The UVP is based on the pulsed Doppler method, and it can measure instantaneous velocity profile along an ultrasonic beam path. In the original UVP principle, it can measure only one-dimensional velocity measurement. Therefore, we extended the UVP to multi-dimensional measurement. To achieve this, a multi-element sensor was used and an algorithm of three-dimensional (3D) velocity vector reconstruction was developed. The 3D vector measurement is realized by simultaneous receiving Doppler signal at each ultrasonic element. The measurement performance was evaluated by the rigid rotating flow measurement, and the relative error of velocity magnitude and vector angle were minimum –1.6% and 11.2% respectively. In addition, we checked the validity of the system for leakage detection by measuring the simulated leakage flow.
Keywords: ultrasound, velocity profile, vector measurement, array transducer, multi-dimensional.

1 INTRODUCTION

Recently, optical inspections have been implemented for the decommissioning of the Fukushima Dai-ichi nuclear power plant (1F). The objective of these inspections is to assess the conditions within the primary containment vessels (PCVs) of units 1, 2 and 3 at the site, and some achievements has been made so far [1]. However, these inspections have not yet unveiled completely the locations of leaks (the repair of which is vital for fuel debris removal) and accurate distribution of fuel debris (an important factor in deciding the future fuel removal procedure). These inspections are hindered by a high-dose radioactive environment and an opaque environment which becomes from the suspended particulates in the coolant water. Therefore, methods other than optical methods are required to inspect within the PCVs.

Ultrasonic measurement is considered as a promising non-optical inspection method. Ultrasonic sound can be used in opaque liquids and ultrasonic transducers are generally suited to high radiation levels, as used in the decommissioning of Three Mile Island (TMI-2) [2]. In our work, an ultrasonic velocity profiler (UVP) [3] and an ultrasonic phased array sensor were used in combination to identify leakage points [4]. The combination system of UVP and phased array sensor allows for the measurement of two-dimensional (2D) flow velocity vector fields using the Doppler frequency shift of echoes scattered by particles in the liquid.

WIT Transactions on Engineering Sciences, Vol 132, © 2021 WIT Press
www.witpress.com, ISSN 1743-3533 (on-line)
doi:10.2495/MPF210131

Hence, leakage points may be identified by observing behaviour of liquid flow near pipes or walls.

However, real flow is three-dimensional (3D) flow, and measurement system is also required to extend to 3D flow measurement. Peronneau et al. [5] proposed a single element cross beam system using two transducers as a transceiver (transmitter/receiver) to measure 2D at the cross point. Further development of a similar measurement system for measuring 3D by using three transducers was shown in the work of Fox [6]. However, this system is time-consuming since the transducers must be operated separately to avoid the interference of the sound beam. Later, Dunmire et al. [7] developed a 3D measurement system using five transducers (one transmitter and four receivers). With only one transmitter, the measurement occurs at the same time and same measurement volume. In fluid engineering, the area of investigation is wider, therefore depth varying (profile) measurement is necessary. Like Dunmire et al. measurement system, Huther and Lemmin [8] developed three-dimensional with varying depth measurement system in open-channel flow. Based on this idea, Obayashi et al. [9] investigated this system accuracy in rotating cylinder flow. They found that the velocity in receiver line has a relatively high error with the reason of low signal to noise ratio.

These studies are very important to be continuously improved since the flow in fluid engineering often exists with multi-dimensional velocity such as 1F case. Owen et al. designed five elements transducer (one transmitter and four receivers) and constructed 3D velocity vector measurement system [10]. This system achieved 3D velocity vector profile measurement with a compact sensor. However, this system reconstructed 3D velocity vector by synthesizing dual 2D velocity vector, and it used large hardware system due to multi element using. In principle, 3D velocity vector can be reconstructed at least one transmitter and three receivers, and it can be more compact system. Furthermore, measurement hardware also can be optimized to the multi elements UVP system.

The purpose of this study is development of the 3D flow vector measurement system with four elements sensor array and optimized hardware system for multi elements UVP measurement. In this paper, the vector UVP system is described, and the velocity profile measurement performance was validated with a rigid rotating flow. Moreover, this system applied to a simulated leakage flow to confirm the validity of 1F investigation.

2 METHOD

2.1 Ultrasonic velocity profiler

An ultrasonic velocity profiler (UVP) can obtain instantaneous velocity profiles of a fluid and it is based on a pulsed Doppler method. Fig. 1 shows the schematic diagram of the UVP measurement. An ultrasonic transducer emits ultrasonic pulses to a particle mixed in the target fluid. The Doppler frequency which depends on the particle velocity can be obtained from the reflected echo signals by the frequency analysing. The velocity component of ultrasonic propagation direction, v, can be written as follows:

$$v = \frac{cf_D}{2f_c}.$$

(1)

where f_D, f_c and c represent the Doppler frequency, the ultrasonic basic frequency of the transducer and the speed of sound in the fluid, respectively. By detecting the Doppler frequency at each measurement position along the beam axis, it realizes to measure the velocity distribution of the fluid. The measurement position, x, can be obtained as follows:

$$x = \frac{c\tau}{2},$$

(2)

where τ is the echo time delay from the transmission time. To detect the Doppler frequency, the UVP method employs a pulse repetition method. Namely, repetition pulses are emitted from the transducer to the particles that be mixed in the target fluid; the Doppler frequency is estimated from the reflected echo waves. Each echo signal accompanies the phase shift because the particle is moved by the flow during the pulse repetition emitting. The auto-correlation method [11] is often used to detect the phase shift between the consecutive two echo signals.

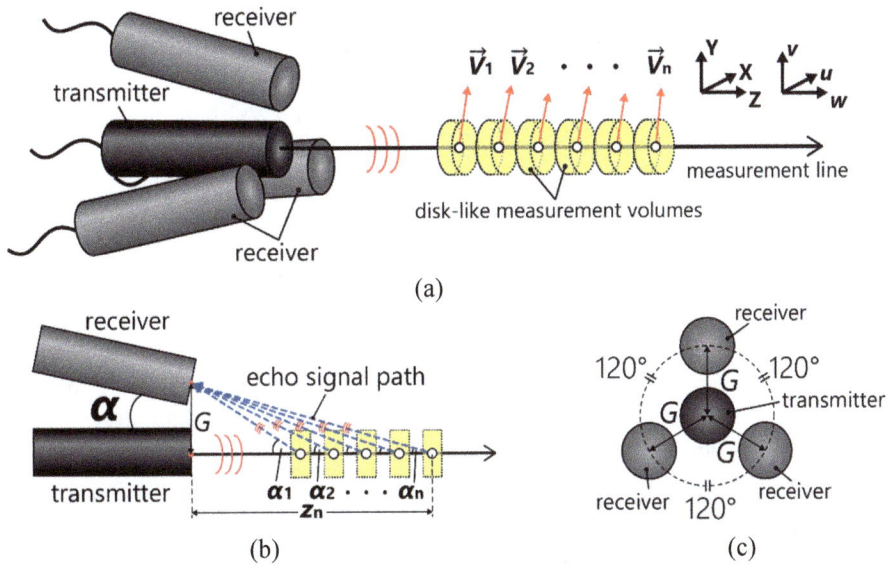

Figure 1: Ultrasonic transducer layout. (a) Overview of vector measurement; (b) Side view of a transmitter-receiver pair; and (c) Front view of the transducer array.

2.2 Three-dimensional velocity vector measurement

Originally, the UVP method obtains one-dimensional velocity profiles (one velocity component on the measurement line) as described at previous section. However, coolant flow structures inside the PCVs are complex due to internal structures and fuel debris; it is needed to extend the UVP to 3D measurement to enhance the efficiency of leak location detection. To extend the velocity component to 3D measurement, some methods have been proposed. Fox et al. [6] proposed a crossbeam method by using three transducers to obtain three velocity components. However, this method is time-consuming since each transducer measurement must be operated separately to avoid the sound beam interference, and it is a spatial point measurement. In contrast, Huther et al. [8] developed 3D velocity vector profile measurement system along on beam line by using five transducers (one transmitter and four receivers). This system obtains the Doppler frequencies from each measurement volume at each receiver simultaneously and reconstructs velocity vector profile; the instantaneous velocity vector profile measurement can be achieved. Using this principle, Owen et al. [10] manufactured the five elements transducer. They achieved 3D velocity vector measurement by one transducer although measurable depth was only approximately 50 mm because the receiver element width was smaller, and the echo sensitivity was decreased.

In this work, to extend the measurable depth and for more compactification of transducer system, the four-transducer array was used. The transducer layout is shown in Fig. 1. The transducer array consists of one transmitter and three receivers (each transducer has a disk shape ultrasonic element). The receivers are equally spaced around the transmitter at 120-degree intervals, with a gap distance G between the transmitter and each receiver. To receive echo signals on the measurement line, each receiver has an angle α between it and the transmitter.

When the tracer particle passing through the n-th measurement volume, three Doppler frequencies, f_{D1n}, f_{D2n} and f_{D3n}, can be obtained from the three receivers. Then, the three velocity components are reconstructed by following equation:

$$
\begin{bmatrix} u_n \\ v_n \\ w_n \end{bmatrix} = \frac{c}{2f_0} \begin{bmatrix} \dfrac{2f_{D1n} - 4\left(f_{D2n} + f_{D3n}\right)}{3\sin\alpha_n} \\[2ex] \dfrac{f_{D2n} - f_{D3n}}{0.866\sin\alpha_n} \\[2ex] \dfrac{f_{D1n} + f_{D2n} + f_{D3n}}{3(1 + \cos\alpha_n)} \end{bmatrix}, \tag{3}
$$

where, c is the speed of sound, f_0 is the basic frequency, and α_n is the angle between the transmitter and receiver at the centre of n-th measurement volume, respectively. The angle α_n is estimated by measurement position:

$$
\alpha_n = \tan^{-1}\left(\frac{G}{z_n}\right), \tag{4}
$$

where, z_n is the z direction distance from the centre transmitter. To detect the Doppler frequency from each receiver, the auto-correlation method [11] was used. The auto-correlation method detect the Doppler phase shift generated by particle moving with quadrature demodulation. In order to reconstruct the velocity vector using eqn (3), the Doppler frequency of the receiver on-axis component should be detected. However, a real Doppler frequency includes other velocity components due to an uncertainty of the echo receivable angle. To suppress the influence of the angle uncertainty and extract the Doppler frequency of on-axis component, beam-axis phase shift was subtracted from detected phase shift at each receiver. The beam-axis phase shift was detected by centre transducer. Namely, the Doppler phase shift, $\Delta\theta_{0i}$ was obtained by following equation:

$$
\Delta\theta_{0i} = \tan^{-1}\left(\frac{\text{Im}\left[\left(s_{0,j} s_{i,j}^* \right)\left(s_{0,j+1} s_{i,j+1}^* \right) \right]}{\text{Re}\left[\left(s_{0,j} s_{i,j}^* \right)\left(s_{0,j+1} s_{i,j+1}^* \right) \right]} \right), \tag{5}
$$

where, subscript of i and j are the receiver index and pulse repetition index, respectively. Character s is the complex signal detected by the quadrature demodulation of the echo signal.

And the mark * represents the conjugate. From this equation, the Doppler frequency of each receiver is estimated by following equation:

$$f_{Di} = \frac{\Delta \theta_{0i}}{2\pi T} ,$$

(6)

where, T is the pulse emission interval time. By using eqns (3)–(6), the velocity vector profile can be reconstructed.

3   VALIDATION OF MEASUREMENT PERFORMANCE

To validate the measurement certainty and accuracy, the rigid rotating flow was measured by the developed system, and the measurement results were compared with theoretical velocity vectors.

3.1 Experimental apparatus

The experimental apparatus for the rigid rotating flow measurement is shown in Fig. 2. As the fluid, tap water was packed in the acrylic cylinder (inner diameter was 154 mm, and wall thickness was 3 mm), and nylon particles were mixed in the water as the tracer particle. The cylinder was rotated by the stepping motor to generate rotating flow inside the cylinder. The rotation speed was controlled by the personal computer. The transducer array was installed at 10 mm from the centre of the cylinder. In the cylinder, the rotating flow is generated, and the theoretical flow vector can be calculated by following equations:

$$v_{theo} = r_x \omega$$
$$w_{theo} = r_z \omega$$

(7)

where, r_x and r_y are x and y components of distance from the cylinder centre, respectively. To evaluate the measurement performance, the measured velocity vectors are compared with the theoretical vectors calculated by eqn (7). The detail of experimental conditions and measurement conditions are shown in Table 1. In this experiment, the distance between the transmitter and cylinder centre was set to 3 position (–1 mm, 2 mm and 8 mm) to validate the effect of the velocity vector angle.

   To realize the three-dimensional velocity vector profile measurement, the measurement instruments was developed. The photograph of developed instruments is shown in Fig. 3. The measurement system consisted of four-transducer array, laboratory made ultrasonic pulser/receiver, and signal processing personal computer. The pulser/receiver included the ultrasound drive circuit, echo signal amplifier and filter, and A/D converter. In this experiment, 150 V and 4 cycle rectangular wave was applied to the transmitter to generate pulsed ultrasound. Echo signal amplifier was set to 55 dB in each receiver. And the A/D converter recorded the echo signals with a sampling speed of 50 MS/s. The recorded signals were transferred to the computer through a USB 3.0 cable. The signal processing was performed in the computer based on eqns (3)–(6), with the programming software LabVIEW 2019 (National Instruments).

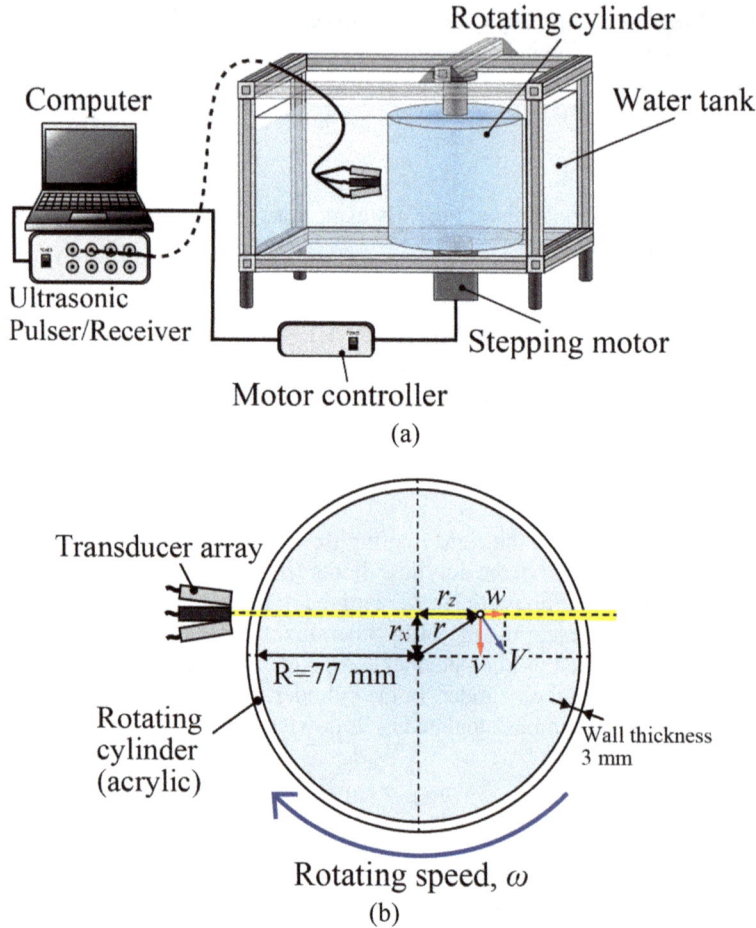

Figure 2:   Experimental setup of the rigid rotating flow measurement. (a) Experimental setup; and (b) Schematic diagram of rigid rotating flow measurement.

Table 1: Experimental conditions of the rigid rotating flow measurement.

Condition and parameter	Detail
Water temperature	25°C
Speed of sound (c)	1,497 m/s
Rotating speed (ω)	600 rpm
Transducer distance between cylinder center (r_x)	−1 mm, 2 mm, 8 mm
Basic frequency (f_0)	4 MHz
Active ultrasonic element diameter	5 mm
Angle between transmitter and receiver	8 degrees
Gap distance (G)	8 mm
Spatial resolution	0.75 mm
Time resolution	64 ms
Number of profiles	1,000

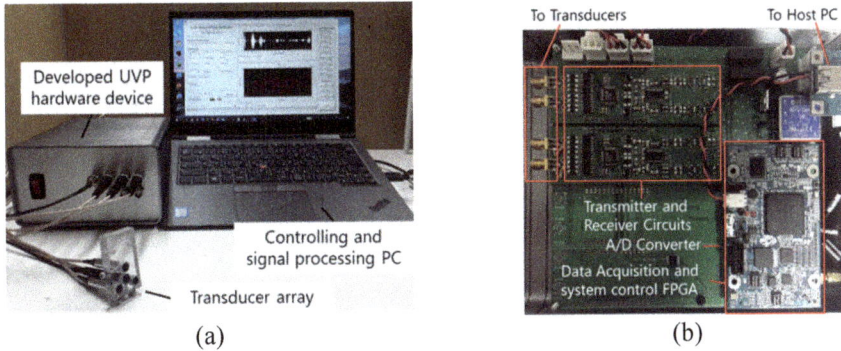

Figure 3: Developed measurement system configuration. (a) Overview of the system; and (b) Internal hardware.

3.2 Results and discussion

The comparison of velocity vector profile between measured and theoretical values are shown in Fig. 4. Fig. 4(a) is the result in the condition of $r_x = -1$ mm, (b) is $r_x = 2$ mm, and (c) is $r_x = 8$ mm position, respectively. Each vector profile is mean of a 1,000 dataset. The horizontal axis is the measurement position from the cylinder centre normalized by cylinder radius R. In the region of r/R = −1 to 0, the receivers were difficult to obtain echo signals due to receivable angle limitation. Therefore, the measurement results are indicated only r/R = 0 to 1. Blue arrows mean theoretical vector, and red allows mean measured vector by the developed system. Figs 5 and 6 show the comparison of theoretical and measured velocity magnitude and vector angle, respectively. These values were calculated by following eqn (8).

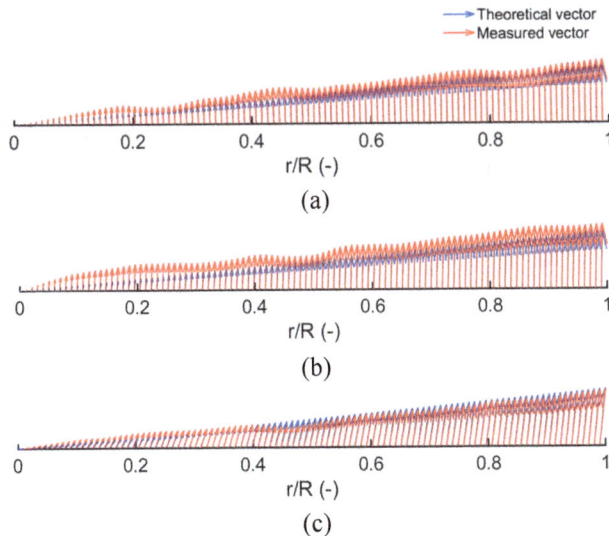

Figure 4: Result of vector profile measurement in the rotating flow. (a) $r_x = -1$ mm; (b) $r_x = 2$ mm; and (c) $r_x = 8$ mm.

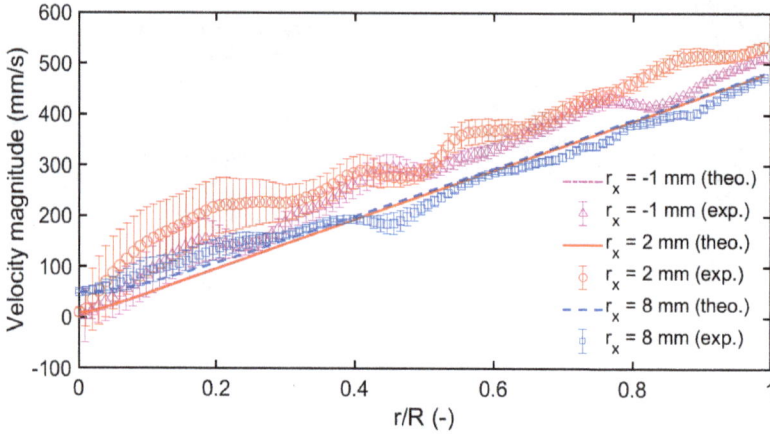

Figure 5: Comparison of theoretical and measured velocity magnitude.

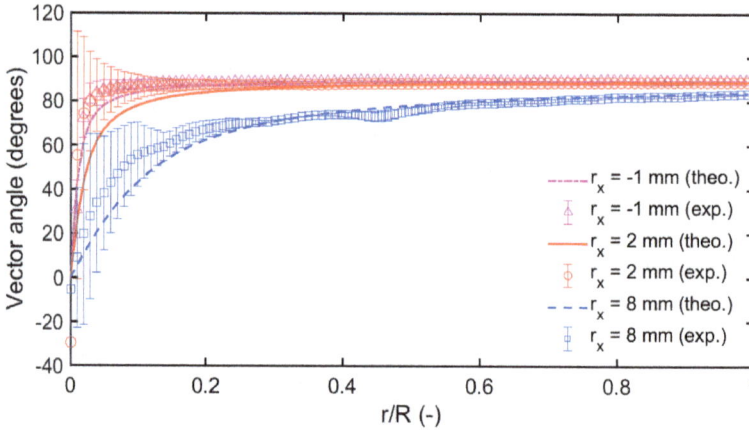

Figure 6: Comparison of theoretical and measured vector angle.

$$|V| = \sqrt{v^2 + w^2}$$
$$\theta = \tan^{-1}\left(\frac{v}{w}\right). \tag{8}$$

In these figures, error bars represent the relative error from the theoretical values. As for the absolute velocity profiles, the spatial averaged relative errors of $r_x = -1$ mm, 2 mm, and 8 mm were –20%, –31% and –1.6%, respectively. And the spatial averaged errors in the vector angle were –24.4%, 16.6%, 11.2%, respectively. Namely, the error increased for vectors that the velocity component was dominated by the lateral component. In this work, the auto-correlation method was used to detect the Doppler frequency at each receiver, however, the method has weakness for very low velocity component. The developed algorithm calculates the phase difference between centre transducer and each receiver to cancel the beam axis Doppler component, but the Doppler component in the beam axis direction was close to zero, which may have increased the error. Nevertheless, each measured vector profile by developed system was captured the rotating flow characteristic.

4 SIMULATED LEAKAGE FLOW MEASUREMENT

To verify the applicability to a leakage detection of the developed system, the simulated leakage flow measurement was performed. We verified whether it is possible to identify the leakage position by scanning the sensor position mechanically and obtaining a three-dimensional flow map.

4.1 Experimental apparatus

Fig. 7 illustrates the experimental setup of the simulated leakage flow measurement. The tap water injected in the water tank (1.2 m × 0.4 m × 0.4 m). The simulated leakage hole was set to the bottom of the water tank, and the hole diameter was 20 mm. The tanked water was

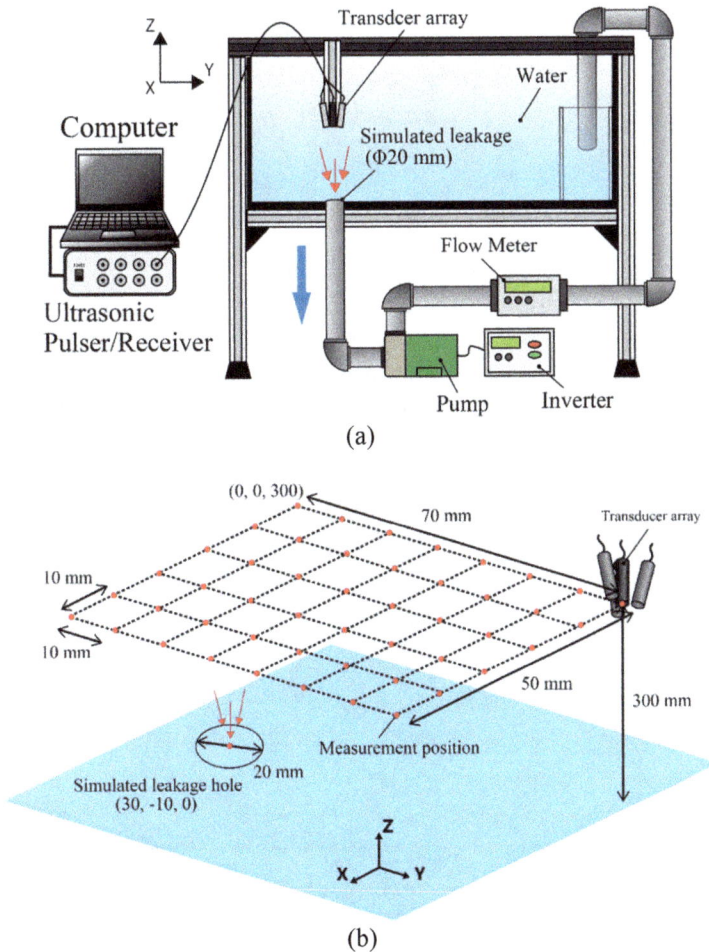

(a)

(b)

Figure 7: Experimental apparatus of simulated leakage flow measurement.
(a) Experimental apparatus; and (b) Schematic diagram of measurement position.

Table 2: Experimental conditions of the simulated leakage flow measurement.

Condition and parameter	Detail
Water temperature	25°C
Speed of sound (c)	1,497 m/s
Leakage hole diameter	20 mm
Leak flowrate	7 L/min
Basic frequency (f_0)	4 MHz
Active ultrasonic element diameter	5 mm
Angle between transmitter and receiver	8 degrees
Gap distance (G)	8 mm
Spatial resolution	0.75 mm
Time resolution	128 ms
Number of profiles at each position	100

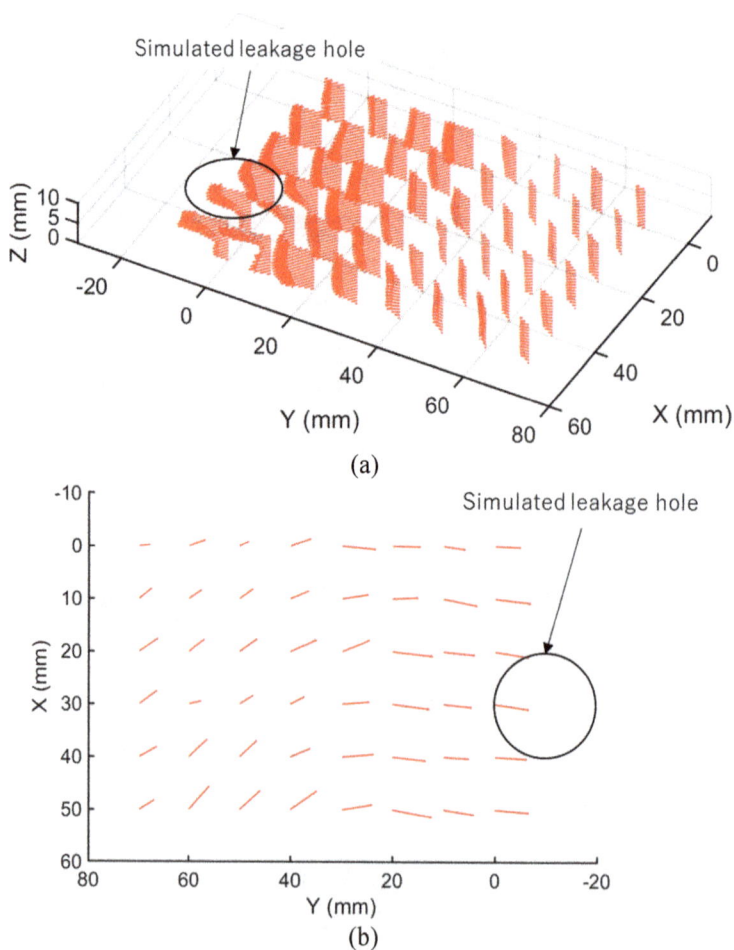

Figure 8: Result of the three-dimensional flow mapping of the simulated leakage flow measurement. (a) 3D view; and (b) xy-plane view.

circulated at 7 L/min from the hole. In this experiment, the transducer array was installed at 300 mm from the bottom of tank, and to obtain the three-dimensional flow map, the transducer array was moved to x and y direction. The measurement position was changed in 50 mm × 70 mm area and the moving pitch distance was set to 10 mm. The centre of leakage hole was (x, y, z) = (30, –10, 0), where the z is the distance from the bottom of tank. The detail of the experimental condition is shown in Table 2. The measurement instrument and conditions were almost same with previous section.

4.2 Results and discussion

The experimental result of simulated leakage flow measurement is shown in Fig. 8. Fig. 8(a) is the three-dimensional vector map around the leakage hole, and (b) is the xy-plane view of it at z = 0 to 10 mm. Where the Fig. 8(b), each velocity vectors are averaged in the region of z = 0 to 10 mm. at the far from the leakage hole, the velocity was low. On the other hand, the velocity vectors with higher velocities toward the simulated leakage hole were observed by developed system. Namely, it was shown that tracking the velocity vector during the flow mapping process can be utilized to identify the location of leakages.

5 CONCLUSION

To detect the coolant water leak positions of Fukushima-Daiichi nuclear power plant, the 3D velocity vector measurement system was developed by using four ultrasonic transducers array and the ultrasonic velocity profiler principle. The velocity vectors were reconstructed with the simultaneous obtained Doppler frequencies at each receiver by the developed algorithm. The measurement performance was evaluated by the rigid rotating flow measurement, and the relative error of velocity magnitude and vector angle were minimum –1.6% and 11.2% respectively. Moreover, the applicability to detection of leakage position was verified by the simulated leakage flow measurement. As the result, it was shown that tracking the velocity vector during the flow mapping process can be utilized to identify the location of leakages. In future work, we will improve the measurement accuracy and contribute to the identification of leakage position in 1F.

ACKNOWLEDGEMENT

This work was supported by Grant-in-Aid for JSPS Fellows Grant Number 20J15257.

REFERENCES

[1]   International Research Institute for Nuclear Decommissioning (IRID), Current status of research and development for decommissioning of Fukushima Daiichi Nuclear Power Plant at IRID. https://irid.or.jp/wp-content/uploads/2019/01/20190108.pdf. Accessed on: 1 May 2021.

[2]   Beller, L.S., Design and operation of the core topography data acquisition system for TMI-2. Three Mile Island Reports. *GEND-INF*-012, 1984.

[3]   Takeda, Y., Development of an ultrasound velocity profile monitor. *Nuclear Engineering and Design*, **126**, pp. 277–284, 1991.

[4]   Nishiwaki, R., Hamdani, A., Takahashi, H., Endo, G. & Kikura, H., Development of a remote water leakage localization system combined with phased array UVP and robot. *Proceedings of 11th International Symposium on Ultrasonic Doppler Methods for Fluid Mechanics and Fluid Engineering*, pp. 88–91, 2018.

[5]   Peronneau, P., Blood flow patterns in large arteries, *Ultrasound in Medicine*, pp. 1193–1208, 1977.

[6]   Fox, M.D. & Gardiner, M.W., Three-dimensional Doppler velocimetry of flow jets. *IEEE Transactions of Biomedical Engineering*, **35**, pp. 834–841, 1988.

[7]   Dunmire, B.L., A vector Doppler ultrasound instrument. *Ultrasonics Symposium, Proceedings*, pp. 1477–1480, 1995.

[8]   Huther, D. & Lemmin, U., A constant-beam-width transducer for 3D acoustic Doppler profile measurements in open-channel flows. *Measurement Science and Technology*, **9**, pp. 1706–1714, 1998.

[9]   Obayashi, H., Tasaka, Y., Kon, S. & Takeda, Y., Velocity vector profile measurement using multiple ultrasonic transducers. *Flow Measurement and Instrumentation*, **19**, pp. 189–195, 2008.

[10]   Owen, J.T., Shoji, N., Hamdani, A., Ihara, T., Takahashi, H. & Kikura., H., Development of new ultrasonic transducer for multi-dimensional velocity profile measurement using ultrasonic Doppler method. *Proceedings of 11th International Symposium on Ultrasonic Doppler Methods for Fluid Mechanics and Fluid Engineering*, pp. 40–43, 2018.

[11]   Kasai, C., Namekawa, K., Koyano, A. & Omoto, R., Real-time two-dimensional blood flow imaging using an autocorrelation technique. *IEEE Transactions on Sonics and Ultrasonics*, **32**, pp. 458–464, 1985.

ACCURATE CONSTITUTIVE RELATIONS FOR SHOCK WAVE STRUCTURES IN GASES

M. H. LAKSHMINARAYANA REDDY & S. KOKOU DADZIE
School of Engineering and Physical Sciences, Heriot-Watt University, Scotland, UK

ABSTRACT

Predicting accurately shock wave structures in gases using an appropriate hydrodynamic model is still a challenge in extended hydrodynamic model development in rarefied gases. In this paper, we identified constitutive equations that provide better agreement for the prediction of shock structure in a monatomic gas in the Mach number range 1.3–4. The results obtained show an improvement upon those obtained previously in the bi-velocity hydrodynamics and are more accurate than in the hydrodynamic models from expansions method solutions to the Boltzmann equation.

Keywords: shock waves, compressible flows, rarefied gas flows, constitutive relations, Navier–Stokes equations, volume/mass diffusion.

1 INTRODUCTION

Shock wave structure description is one of the best-known example of a simple highly non-equilibrium compressible flow problem where large gradients of hydrodynamic fields are present [1]–[8]. Shock structure problem serves as a standard benchmark problem for testing the capability (validity) and accuracy of different hydrodynamics and extended hydrodynamic fluid flow models and provides few advantages in its numerical simulations: (i) flow is one-dimensional and steady state; (ii) no solid boundaries; and (iii) the upstream and downstream states are in equilibrium and are connected by simple relations (the Rankine–Hugoniot relations). Theoretical and numerical studies of the shock structure based on the classical Navier–Stokes (NS) equations are described in the literature [1], [2], [9]–[12]. In addition, shock density measurements have been carried out and reported for argon and nitrogen gases with Mach number that ranges from supersonic to hypersonic [13], [14]. Eventually, it has been recognized that shock structures in monatomic gases are not well described by the Navier–Stokes theory. The shock thickness predicted is too small compared to experiments for Mach numbers larger than approximately 1.5 (see. Fig. 1). The failure of the equations in a shock structure description may be tied up to the basic assumptions such as linear constitutive relations represented by Newton's law of viscosity and Fourier's law of heat conduction used in closing the system [15] and/or breakdown of continuum assumption as the mean free path becomes comparable to the characteristic length scale of the system.

The principal parameter which is often used to classify the non-equilibrium state of a gas flow is the Knudsen number, Kn, and is defined as the ratio of the mean free path of the gas molecules to the characteristic length of the flow system. Kn characterizes the gas rarefaction which means that it measures departure from the local equilibrium. Continuum assumption is valid for vanishing Knudsen numbers where the gas can be assumed to undergo a large number of collisions over the typical length scale. As Kn increases the notion of the gas as a continuum fluid becomes less valid and the departure of the gas from the local thermodynamic equilibrium increases. Therefore, the range of use of the continuum-equilibrium assumption is limited and confined to $\mathrm{Kn} \lesssim 0.01$. Generally, the shock macroscopic parameter called the inverse shock thickness is related to the Knudsen number and typically falls between ≈ 0.2 and ≈ 0.3 [6] (which is also evident from Fig. 1). Clearly, the range of Kn found in the shock problem is beyond the classical continuum-Kn regime and falls into the so-called

WIT Transactions on Engineering Sciences, Vol 132, © 2021 WIT Press
www.witpress.com, ISSN 1743-3533 (on-line)
doi:10.2495/MPF210141

'intermediate-Kn' regime ($0.01 \lesssim \mathrm{Kn} \lesssim 1$). Deriving appropriate continuum hydrodynamic models or improving the range of applicability of the existing ones (the Navier–Stokes equations) beyond their limits into the so-called 'intermediate-Kn' regime ($0.01 \lesssim \mathrm{Kn} \lesssim 1$) is still a critical area of research.

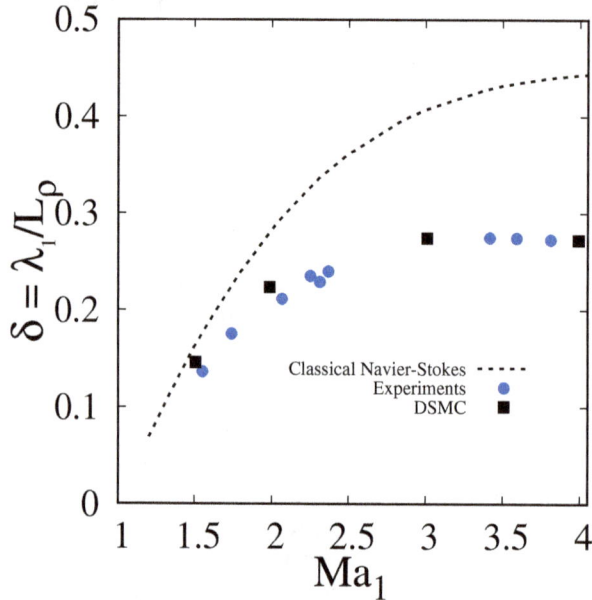

Figure 1: Variation of dimensionless inverse shock thickness (δ) verses upstream Mach number $\mathrm{Ma_1}$ for a monatomic argon gas. Experimental data extracted from Alsmeyer [13] and DSMC data extracted from Lumpkin and Chapman [16].

Recently, the authors reinterpreted shock structure predictions of the classical Navier–Stokes equations using a change of velocity variable [17]–[19]. The results on the shock density profiles and shock thicknesses better agreed with the experimental data. However, the procedure predicted very less values for density asymmetry factor. The present work expands on this previous work to identify constitutive relations with a full assessment of the shock structure problem.

The paper is organized as follows. In Section 2 a brief overview of the classical Navier–Stokes equations for compressible flows and the modified constitutive relations are presented. In Section 3 the modified Navier–Stokes equations are considered subject to shock structure problem in monatomic argon gas. The detail of the formulation of the problem and numerical procedure are then provided. Section 4 is completely devoted to analysis based on comparison of shock macroscopic profiles and different shock macroscopic parameters with available experiments and other simulation data. Finally, conclusions are drawn in Section 5.

2   THE CONTINUUM FLOW EQUATIONS: MODIFIED CONSTITUTIVE RELATIONS

The standard continuum hydrodynamic equations are a differential form of three classical conservation laws, namely, mass, momentum and energy conservation laws that govern the motion of a fluid. Here, we adopt the classical conservation equations in an Eulerian reference frame as given by:

mass balance equation

$$\frac{\partial \rho}{\partial t} + \nabla \cdot [\rho \, \mathrm{U}] = 0, \tag{1}$$

momentum balance equation

$$\frac{\partial \rho \, \mathrm{U}}{\partial t} + \nabla \cdot [\rho \, \mathrm{U} \otimes \mathrm{U} + \mathrm{p} \, \mathbf{I} + \mathbf{\Pi}] = 0, \tag{2}$$

energy balance equation

$$\frac{\partial}{\partial t} [\frac{1}{2} \rho \, \mathrm{U}^2 + \rho \, \mathbf{e}_{\mathrm{in}}] + \nabla \cdot [\frac{1}{2} \rho \, \mathrm{U}^2 \, \mathrm{U} + \rho \, \mathbf{e}_{\mathrm{in}} \, \mathrm{U} + (\mathrm{p} \, \mathbf{I} + \mathbf{\Pi}) \cdot \mathrm{U} + \mathbf{q}] = 0, \tag{3}$$

where ρ is the mass-density of the fluid, U is the flow mass velocity, p is the hydrostatic pressure, $\mathbf{e}_{\mathrm{in}}$ is the specific internal energy of the fluid, $\mathbf{\Pi}$ is the shear stress tensor, $\mathbf{I}$ is the identity tensor and $\mathbf{q}$ is the heat flux vector. All these hydrodynamic fields are functions of time t and spatial variable $\mathbf{x}$. Additionally, ∇ and $\nabla\cdot$ denote the usual spatial gradient and divergence operators, respectively, and $\mathrm{U} \otimes \mathrm{U}$ represents the tensor product of two velocity vectors. Expression for the specific internal energy is given by, $\mathbf{e}_{\mathrm{in}} = \mathrm{p}/\rho(\gamma - 1)$ with γ being the isentropic exponent. The constitutive models for the shear stress $\mathbf{\Pi}$ and the heat flux vector $\mathbf{q}$ as due to the Newton's law and the Fourier's law, are given respectively by,

$$\mathbf{\Pi}^{(\mathrm{NS})} = -2 \, \mu \left[\frac{1}{2} (\nabla \mathrm{U} + \nabla \mathrm{U}') - \frac{1}{3} \mathbf{I} \, (\nabla \cdot \mathrm{U}) \right] = -2 \, \mu \, \overset{\circ}{\overline{\nabla \mathrm{U}}}, \quad \mathbf{q}^{(\mathrm{NS})} = -\kappa \, \nabla \mathrm{T}, \tag{4}$$

where $\nabla \mathrm{U}'$ represents the transpose of $\nabla \mathrm{U}$. Coefficients μ and κ are the dynamic viscosity and the heat conductivity, respectively.

The eqns (1)–(4) is the well known and widely accepted conventional fluid flow hydrodynamic model for a viscous and heat conducting fluid. In the limit of vanishing viscous and heat conducting terms, the model reduced to the simple gas dynamics model known as Euler equations, which are used to model inviscid and non-diffusive flows. In the present study, to investigate our shock structure problem we adapt constitutive equations from previous studies (Dadzie [20], Brenner [21]):

$$\mathbf{\Pi} = -2 \, \mu \, \overset{\circ}{\overline{\nabla \mathrm{U}}} - 2 \, \mu \, \overset{\circ}{\overline{\nabla \mathrm{J}_{\mathrm{D}}}}, \quad \mathbf{q} = -\kappa \, \nabla \mathrm{T} - \frac{\gamma}{(\gamma - 1)\mathrm{Pr}} \, \mathrm{p} \, \mathrm{J}_{\mathrm{D}}, \tag{5}$$

where Pr is the Prandtl number and is a dimensionless number defined as the ratio of momentum diffusivity to thermal diffusivity, $\mathrm{J}_{\mathrm{D}} = \kappa_{\mathrm{m}} \nabla \ln \rho$ and κ_{m} is an additional transport coefficient, the molecular diffusivity coefficient and is here related to the kinematic viscosity coefficient through the following relation:

$$\kappa_{\mathrm{m}} = \kappa_{\mathrm{m}_0} \frac{\mu}{\rho}, \qquad \text{where } \kappa_{\mathrm{m}_0} \text{ is a positive constant.} \tag{6}$$

It is note worthy to point here that constitutive relations given in eqn (5) are formally those proposed in Dadzie [20], Brenner [21] and used in Greenshields and Reese [6]. The only main difference in the current expression being the additional factor of $\gamma/((\gamma - 1)\mathrm{Pr})$ in the heat flux relation. In the following sections, we reveal that these constitutive relations show the best prediction of the shock profile in a monatomic argon gas.

3 THE SHOCK STRUCTURE PROBLEM IN A MONATOMIC GAS: FORMULATION AND NUMERICAL PROCEDURE

The evolution of a monatomic ideal gas flow is determined by the density ρ, the velocity U and the temperature T at any point in space and time. Its pressure p obeys the perfect gas law,

$$p = \rho \mathbf{R} T, \tag{7}$$

where $\mathbf{R} = k_B/m$ is the specific gas constant with k_B and m being the Boltzmann constant and the molecular mass, respectively. In terms of the specific heat at constant pressure, c_p, and constant volume, c_v, a monatomic ideal gas is characterized by,

$$c_p = \frac{\gamma}{(\gamma - 1)} \mathbf{R}, \qquad c_v = \frac{1}{(\gamma - 1)} \mathbf{R}, \tag{8}$$

such that the ratio of c_p to c_v, called the isentropic constant γ, is equal to $5/3$.

We consider a planar shock wave propagating in the positive x-direction which is established in a flow of a monatomic gas. For this one-dimensional flow problem, all hydrodynamic variables are functions of a single spatial coordinate x and time t; the system is assumed to be uniform (having no gradients) and infinite along the y- and z-directions. The flow velocity and heat flux in the x-direction are denoted by $u(x, t)$ and $q(x, t)$, respectively, and are zero in the two remaining (y and z) orthogonal directions. It is straightforward to verify that the stress tensor has only one non-zero component, the longitudinal stress which can be expressed as,

$$\Pi_{xx} = -\frac{4}{3}\mu \frac{\partial u}{\partial x} - \frac{4}{3}\frac{\mu \kappa_m}{\rho}\frac{\partial^2 \rho}{\partial x^2} + \frac{4}{3}\frac{\mu \kappa_m}{\rho^2}\left(\frac{\partial \rho}{\partial x}\right)^2 \equiv \Pi, \tag{9}$$

and the constitutive relation for the heat flux is,

$$q = -\kappa \frac{\partial T}{\partial x} - \frac{c_p}{Pr}\kappa_m \rho T \frac{\partial \ln \rho}{\partial x}. \tag{10}$$

With the above definitions, the one-dimensional reduced balance equations for the modified Navier–Stokes model can be written in 'conservative' form:

$$\frac{\partial \rho}{\partial t} + \frac{\partial}{\partial x}(\rho u) = 0, \tag{11}$$

$$\frac{\partial}{\partial t}\left(\rho u\right) + \frac{\partial}{\partial x}\left(\rho u^2 + \rho \mathbf{R} T + \Pi\right) = 0, \tag{12}$$

$$\frac{\partial}{\partial t}\left(\frac{1}{2}\rho u^2 + c_v T\right) + \frac{\partial}{\partial x}\left(\frac{1}{2}\rho u^3 + c_p \rho T u + \Pi u + q\right) = 0. \tag{13}$$

The one-dimensional classical Navier–Stokes system is obtained by setting $\kappa_m = 0$ in the constitutive relations of longitudinal stress and heat flux, i.e., in eqns (9) and (10), respectively. The corresponding Euler system is then obtained by setting $\Pi = 0$ and $q = 0$ in eqns (11)–(13).

We denote the upstream $(x \to -\infty)$ and downstream $(x \to \infty)$ conditions of a shock, located at $x = 0$, by a subscript 1 and 2, respectively. That is the upstream and the down stream equilibrium states are characterised by (ρ_1, u_1, T_1) and (ρ_2, u_2, T_2), respectively. Across a shock, the finite jump in each state variable is given by the so-called Rankine–Hugoniot (RH) relations [1], [2] that connect the upstream and downstream states of a shock.

These relations can be obtained from the conservation balance laws (11)–(13) by following the standard procedure given in [1] and employing the ideal gas equation of state. The standard Rankine–Hugoniot relations can be obtained as:

$$\rho_1\,u_1 = \rho_2\,u_2,$$
$$\rho_1\,u_1^2 + \rho_1\,\mathbf{R}\,T_1 = \rho_2\,u_2^2 + \rho_2\,\mathbf{R}\,T_2, \tag{14}$$
$$\rho_1\,u_1^3 + 2\,c_p\,\rho_1\,T_1\,u_1 = \rho_2\,u_2^3 + 2\,c_p\,\rho_2\,T_2\,u_2.$$

The modified Navier–Stokes equations, for the one-dimensional stationary shock flow configuration reduced to:

$$\frac{d}{dx}\left[\rho\,u\right] = 0, \qquad \frac{d}{dx}\left[\rho\,u^2 + p + \Pi\right] = 0, \tag{15}$$

$$\frac{d}{dx}\left[\rho\,u\left(\frac{1}{2}u^2 + c_p\,T\right) + \Pi u + q\right] = 0. \tag{16}$$

Integration of the eqns (15)–(16) leads to:

$$\rho\,u = m_0, \qquad \rho\,u + \rho\,\mathbf{R}\,T + \Pi = p_0, \tag{17}$$

$$\rho\,u\left(c_p\,T + \frac{u^2}{2}\right) + \Pi u + q = m_0\,h_0, \tag{18}$$

where m_0, p_0 and h_0 are integration constants which represents the mass flow rate, the stagnation pressure and the stagnation specific enthalpy, respectively, and their values can be obtained using the well-known Rankine–Hugoniot conditions (14). In order to solve the eqns (17) and (18), it is convenient to work with its dimensionless form. We use the following set of dimensionless variables based on the upstream reference states (denoted with subscript 1) as in Reese et al. [5], Reddy and Dadzie [17], and Dadzie and Reddy [22]:

$$\overline{\rho} = \frac{c_1^2}{p_1}\rho = \frac{\gamma}{\rho_1}\rho, \quad \overline{u} = \frac{u}{c_1}, \quad \overline{T} = \frac{\mathbf{R}}{c_1^2}T, \quad \overline{p} = \frac{p}{p_1}, \quad \overline{x} = \frac{x}{\lambda_1}, \quad \overline{\mu} = \frac{\mu}{\mu_1}, \tag{19}$$

where λ_1 is the upstream mean free path which is a natural choice for a characteristic length-scale as changes through the shock occur due to few collisions and $c_1 = \sqrt{\gamma\,\mathbf{R}\,T_1}$ being the adiabatic sound speed. The upstream mean free path can be expressed as a function of reference state variables: $\lambda_1 = \lambda_0\mu_1/\rho_1\,c_1$, with $\lambda_0 = (16/5)\sqrt{2\,\pi/\gamma}$. Dimensionless forms of transport coefficients $\overline{\kappa}$ and $\overline{\kappa}_m$ are:

$$\overline{\kappa} = \frac{\gamma}{(\gamma-1)\,\mathrm{Pr}}\overline{\mu} \quad \text{and} \quad \overline{\kappa}_m = \kappa_{m_0}\frac{\overline{\mu}}{\overline{\rho}}, \tag{20}$$

with the Prandtl number, Pr, equals to $2/3$ for the case of a monatomic gas.

It is well-known that the viscosity and temperature relation has a noticeable effect on the shock wave structure. Here we adopt the generally accepted temperature-dependent viscosity power law [6], [16]: $\mu \propto T^s$ or $\mu = \alpha\,T^s$, where α is a constant of proportionality taken to be γ^s and the power s for almost all real gases falling between $0.5 \leq s \leq 1$, with the limiting cases, $s = 0.5$ and $s = 1$ corresponding to theoretical gases, namely, the hard-sphere and Maxwellian gases, respectively. In our simulations we use $s = 0.75$ for a monatomic argon gas.

The nondimensionalized form of the integral conservation eqns (17) and (18) can then be obtained using the dimensionless quantities defined via eqns (19) and (20) as:

$$\bar{\rho}\,\bar{u} = \bar{m}_0, \qquad -\frac{1}{\lambda_0\,\mathrm{Ma}_1}\overline{\Pi} = \frac{\overline{T}}{\bar{u}} + \bar{u} - \bar{p}_0, \tag{21}$$

$$-\frac{(\gamma-1)}{\lambda_0\,\mathrm{Ma}_1}\bar{q} = \overline{T} - \frac{(\gamma-1)}{2}\bar{u}^2 + (\gamma-1)\,\bar{p}_0\,\bar{u} - \bar{h}_0, \tag{22}$$

where Ma_1 is the upstream Mach number defined as the ratio of the speed of the gas to the speed of sound through the gas, $\mathrm{Ma}_1 = u_1/c_1$. Expressions for the quantities $\bar{m}_0$, $\bar{p}_0$ and $\bar{h}_0$ can be then obtained as,

$$\bar{m}_0 = \gamma\,\mathrm{Ma}_1, \quad \bar{p}_0 = \frac{1}{\gamma\,\mathrm{Ma}_1}\left(1 + \gamma\,\mathrm{Ma}_1^2\right), \quad \bar{h}_0 = 1 + \frac{(\gamma-1)}{2}\mathrm{Ma}_1^2, \tag{23}$$

and the expressions for the dimensionless shear stress ($\overline{\pi}$) and the heat flux ($\bar{q}$) are given by,

$$\overline{\Pi} = -\frac{4}{3}\bar{\mu}\frac{d\bar{u}}{d\bar{x}} - \frac{4}{3}\left(\frac{\gamma}{\lambda_0}\right)\frac{\bar{\mu}\,\bar{\kappa}_m}{\bar{\rho}}\frac{d^2\bar{\rho}}{d\bar{x}^2} + \frac{4}{3}\left(\frac{\gamma}{\lambda_0}\right)\frac{\bar{\mu}\,\bar{\kappa}_m}{\bar{\rho}^2}\left(\frac{d\bar{\rho}}{d\bar{x}}\right)^2, \tag{24}$$

$$\bar{q} = -\bar{\kappa}\frac{d\overline{T}}{d\bar{x}} - \frac{\gamma}{(\gamma-1)\,\mathrm{Pr}}\bar{\kappa}_m\,\overline{T}\frac{d\bar{\rho}}{d\bar{x}}. \tag{25}$$

We solve the final eqns (21) and (22) using a numerical scheme, namely, finite difference global solution (FDGS) technique developed by Reese et al. [5] with well-posed boundary conditions. For the specific details of FDGS scheme reader can refer to Reese et al. [5].

4 NUMERICAL RESULTS AND COMPARISON OF HYDRODYNAMIC FIELDS ACROSS THE SHOCK LAYER

We perform numerical simulations of stationary shock waves located at $x = 0$ using FDGS scheme by considering a computational spatial domain of length $100\lambda_1$ covering $(-50\lambda_1, 50\lambda_1)$ with a step size equal to $\approx 0.05\lambda_1$. This domain is wide enough to contain the entire shock profile for weak shocks ($\mathrm{Ma}_1 \sim 1$) without altering its structure in a monatomic argon gas. We assume the constant κ_{m_0} in the molecular mass diffusivity κ_m to be 0.5 in all our present simulations and for this value of κ_{m_0} the system is linearly stable. To compare the shock structure profiles among the theoretical and experimental data, the position x has been scaled such that $x = 0$ corresponds to a value of the normalized gas density $\rho_N = (\rho - \rho_1)/(\rho_2 - \rho_1)$ equals 0.5. Other hydrodynamic field, namely, the velocity profiles are normalised via: $u_N = (u - u_2)/(u_1 - u_2)$. Fig. 2(a) and (b) shows the comparison of the normalised velocity profiles obtained from the classical and the present modified Navier–Stokes equations for $\mathrm{Ma}_1 = 1.55$ and $\mathrm{Ma}_1 = 4$, respectively. As the flow varies from supersonic to subsonic across the shock, the velocity is maximum/high at the upstream part of the shock, decreases through the shock and attains its smallest value at the downstream part of the shock. The velocity profiles obtained from the modified NS model are more diffusive than the classical NS profile at both upstream and downstream part of the shock which is evident from Fig. 2.

4.1 Density profiles across the shock

Experimental data exist for monatomic argon gas density variations across shock layer. We take a detailed comparison of density field across the shock with experimental results

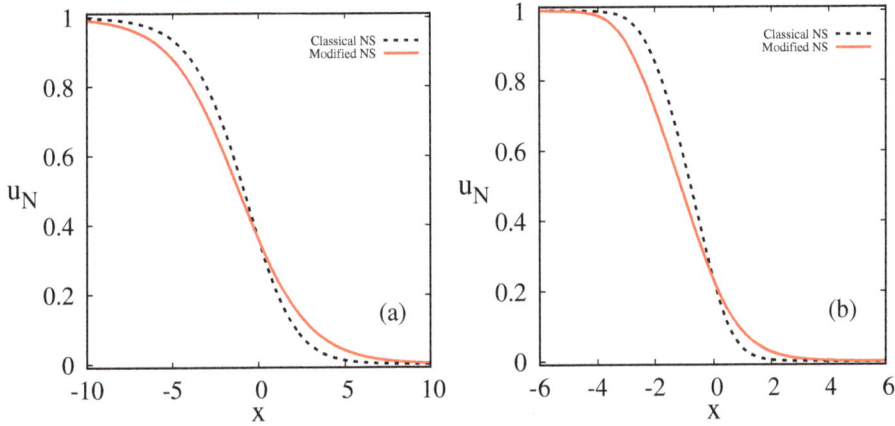

Figure 2: Variation of normalized velocity (u_N) profiles in Ar shock layer: for (a) $Ma_1 = 1.55$; and (b) $Ma_1 = 4$.

of Alsmeyer [13] and also with the classical Navier–Stokes equations. Fig. 3 show the normalized density ρ_N profiles through an argon shock wave as predicted by the modified Navier–Stokes and the classical Navier–Stokes equations compared with the experimentally measured density data. Panels (a), (b), (c) and (d) of Fig. 3 correspond to upstream Mach numbers of $Ma_1 = 1.55, 2.05, 3.38$ and 3.8, respectively. In each panel: the dotted black lines represent solutions of the Navier–Stokes equations and the red lines represent solutions by the modified Navier–Stokes equations. The filled blue circles represent the experimental data of Alsmeyer [13]. From panel (a) of Fig. 3 one observes that for the upstream Mach number of 1.55 the classical Navier–Stokes is able to predict well the upstream part of the shock layer in comparison with the experimental data but completely fails to predicts the downstream part of the shock layer. The modified Navier–Stokes equations produce good agreement with the experimental data with a small disparity at the upstream part of the shock layer and is more diffusive than the experimental data. The modified Navier–Stokes predictions for the normalized density profiles show excellent agreement with the experimental data for the upstream Mach number of $Ma_1 = 2.05, 3.38$ and 3.8, which is evident from panels (b)–(d) of Fig. 3. Overall, an excellent agreement between predictions of the modified Navier–Stokes equations and the experimental data of Alsmeyer [13] is found for weak shocks ($Ma_1 \sim 1$) to moderate strong shocks ($Ma_1 \sim 4$), which is seen in Fig. 3.

4.2 Shock macroscopic parameters: shock thickness and density asymmetry

We discuss in this section two important parameters, namely, shock thickness L_ρ and density asymmetry factor Q_ρ which are often used to characterize the shock wave properties instead of comparing full shock wave profiles. These two shock macroscopic parameters are defined based on shock density profiles. The usual shock thickness or width L_ρ is defined as (Greenshields and Reese [6], Reddy and Alam [7], [8], Gilbarg and Paolucci [10], Pham-Van-Diep et al. [12] and Alsmeyer [13]):

$$L_\rho = \frac{\rho_2 - \rho_1}{|\max(\frac{d\rho}{dx})|}, \tag{26}$$

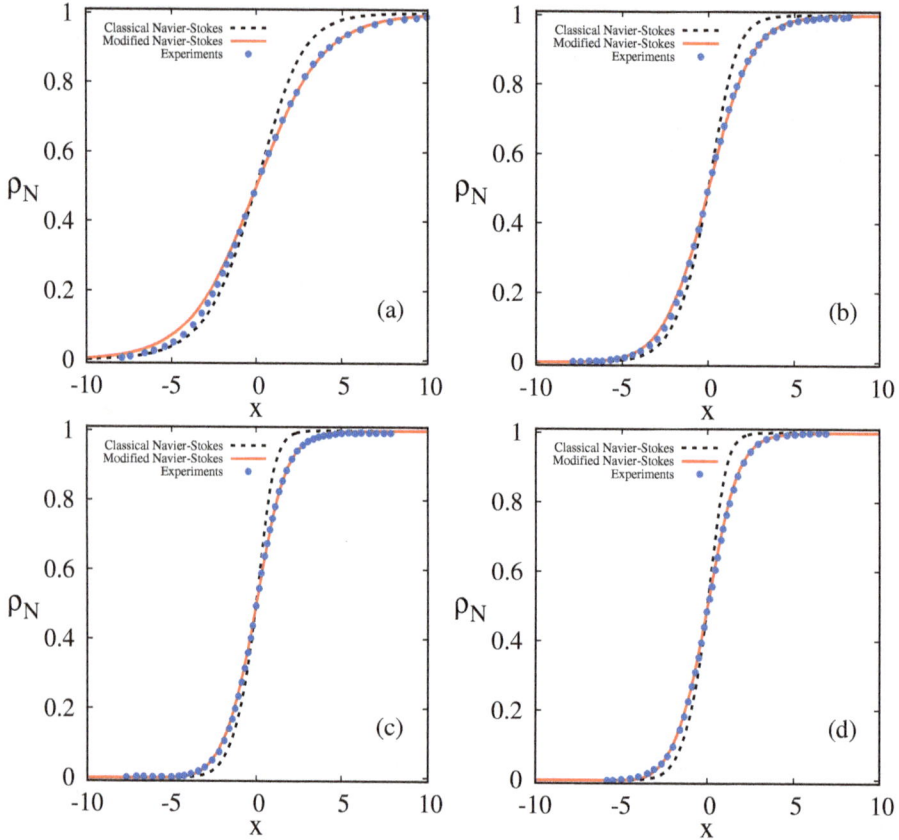

Figure 3: Variation of normalized density (ρ_N) profiles in Ar shock layer: for (a) $Ma_1 = 1.55$; (b) $Ma_1 = 2.05$; (c) $Ma_1 = 3.38$; and (d) $Ma_1 = 3.8$.

and is based on the density profile and depends mainly on the central part of the shock wave. Note that from the definition of L, one can infer that it has a linear dependence on the density difference between the upstream and downstream states and inversely proportional to a slope corresponding to the maximum density gradient. In general, the non-dimensional inverse shock thickness $\delta = \lambda_1/L_\rho$ is used instead of shock thickness L_ρ to compare computational results with experiments as it possesses an important feature that is, it represents actually the Knudsen number of the shock structure flow problem.

In Fig. 4, the prediction of the modified Navier–Stokes equations (red line with filled rhombus) for reciprocal shock thickness (δ) as a function of upstream Mach number (Ma_1) are compared with the theoretical results from Burnett equations [16] (double dot dashed line), a second-order continuum equations of Paolucci and Paolucci [23] (dash dot line) along with the Alsmeyer experimental data (open and filled circles) and also with the DSMC data (filled squares). Predictions from the classical Navier–Stokes with $s = 0.75$ (see black dotted line) are also presented for the sake of completeness. From Fig. 5, one can observe that the classical Navier–Stokes equations with $s = 0.75$ predict the higher reciprocal shock thickness than the measured values over the entire Mach number range presented. In other

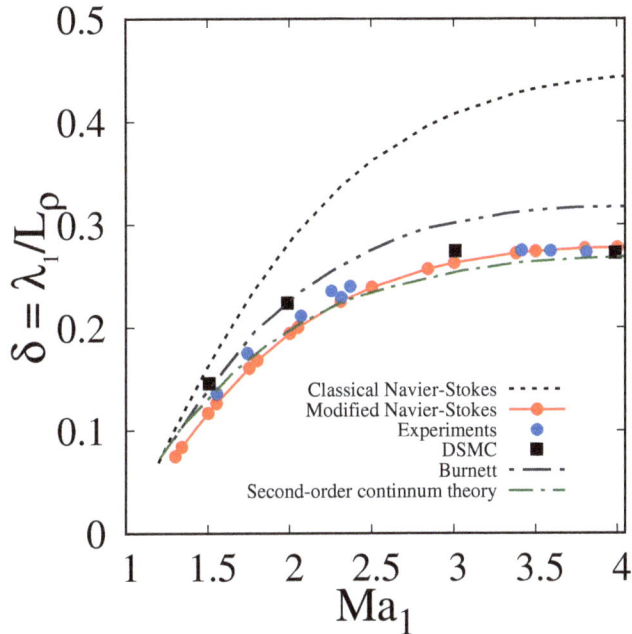

Figure 4: Variation of the dimensionless inverse shock thickness (δ) verses upstream Mach number Ma_1. Line with dot (red) represent modified Navier–Stokes solution with $\kappa_{m_0} = 0.5$ and $s = 0.75$, dashed line (black) shows present solutions of the NS with $s = 0.75$ using FDGS technique, results from Burnett theory and DSMC [16], second-order continuum theory [23] and experiments [13] are superimposed.

words we can say that the classical Navier–Stokes equations predicts very small values for the shock thickness (L_ρ). While the predictions from the modified Navier–Stokes equations with $s = 0.75$ is found to follow very closely the experimental results of Alsmeyer [13] (see filled blue circles). It can be seen that the predictions of the modified NS equations show a close reasonable agreement with the DSMC results of Lumpkin and Chapman [16] at all upstream Mach numbers ranging from $Ma_1 = 1.55$ to 4. Some deviations from the Alsmeyer experimental results are observed in Burnett and second-order continuum model of Paolucci and Paolucci [23] which is evident from Fig. 4. Overall, judged by the inverse shock thickness, it has been found from Fig. 4 that the modified Navier–Stokes model gives good agreement with the experimental results of Alsmeyer [13] and a reasonable good agreement with the DSMC results [16]. Furthermore, the inverse shock thickness predictions of the modified NS equations showed improvement over the Burnett and the second-order continuum models.

A second important shock macroscopic parameter called the density asymmetry factor Q_ρ can be used to describe the actual shape of the shock structure as it measures skewness of the density profile relative to its midpoint [6]. The shock asymmetry, Q_ρ, is defined based on the normalized density profile, ρ_N, with its centre, $\rho_N = 0.5$, located at x = 0, as

$$Q_\rho = \frac{\int_{-\infty}^{0} \rho_N(x)\,dx}{\int_{0}^{\infty} [1 - \rho_N(x)]\,dx}. \tag{27}$$

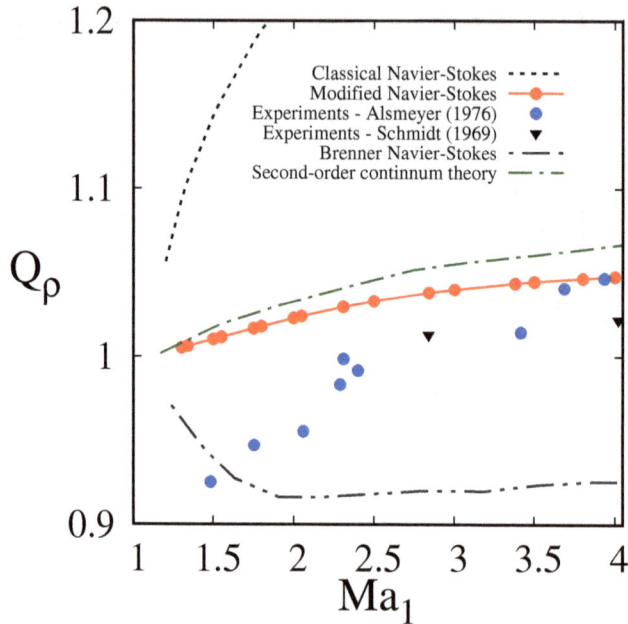

Figure 5: Variation of density asymmetry or shape factor (Q_ρ). Line with dot (red) represent modified Navier–Stokes solution with $\kappa_{m_0} = 0.5$ and $s = 0.75$, dashed line (black) shows present solutions of the NS with $s = 0.75$ using FDGS technique, results from Brenner Navier–Stokes [6], second-order continuum theory [23] and experiments [13], [14] are superimposed.

From eqn (27) it is clear that, a symmetric shock wave profile will have a density asymmetry quotient of unity, while for realistic shock waves its value is around unity and asymmetric shock profiles are predicted in experiments for strong hypersonic shock waves for which Q_ρ is always grater than unity.

Predictions of the modified Navier–Stokes equations for the density asymmetry quotient Q_ρ compared with experimental data of Alsmeyer [13] and Schmidt [14] along with the other theoretical predictions from the classical Navier–Stokes, a second-order continuum theory of Paolucci and Paolucci [23] and Brenner Navier–Stokes equations [6] are shown in Fig. 5. One can observe that the classical Navier–Stokes equations predict an asymmetry quotient of more than unity (which means that the upstream part of the shock profile is more skewed than at the downstream) at all upstream Mach numbers and are far away from the experimental predictions. Brenner Navier–Stokes equations predict $Q_\rho < 1$ for all upstream Mach numbers studied. In contrast to the predictions of classical and Brenner Navier–Stokes, it is experimentally found and reported that the density profile has a significant asymmetry ($Q_\rho = 1 \pm 0.15$) at all upstream Mach numbers. The modified Navier–Stokes predicts an asymmetry quotient of around unity at low upstream Mach numbers and its value increases with shock strength which is evident from Fig. 5. It can be seen that the current modified Navier–Stokes and second-order continuum theory of Paolucci and Paolucci [23] are better in predicting reasonable density asymmetry factor at all upstream Mach numbers range studied here.

5 CONCLUSIONS

In this work we presented a numerical investigation into shock wave profile description in a monatomic gas using hydrodynamics models by identifying constitutive equations that provide better agreement for the parameters involved in the prediction. A detailed comparison between the predictions of the modified hydrodynamic equations with Alsmeyer's experimental data and DSMC data along with classical Navier–Stokes hydrodynamic solutions presented for upstream Mach number range 1.3 to 4. First, we focused our comparison for shock density profiles as accurate data from the experiments are available. Second, we showed the comparison of two well-known shock macroscopic parameters (inverse shock thickness and density asymmetry factor) with available experimental and DSMC data. Our analysis showed that the constitutive equations provide excellent quantitative agreement with experimental data as well as with DSMC data at all Mach numbers discussed and especially best in reproducing experimental trends for the shock density and inverse shock thickness profiles. In fact we conclude that the results are improvement upon those obtained previously in bi-velocity hydrodynamics and showed improvements over those obtained using equations from the extended hydrodynamic approach of kinetic theory. Further implications of these results as related to recently proposed recast Navier–Stokes equations are still to be investigated.

ACKNOWLEDGEMENTS

This research is supported by the UK's Engineering and Physical Sciences Research Council (EPSRC) under grant no. EP/R008027/1 and The Leverhulme Trust, UK, under grant Ref. RPG-2018-174.

REFERENCES

[1] Courant, R. & Friedrichs, K.O., *Supersonic Flow and Shock Waves*. Interscience: New York, 1948.

[2] Liepmann, H.W. & Roshko, A., *Elements of Gas Dynamics*. John Wiley & Sons, Inc.: New York, 1957.

[3] Grad, H., The profile of a steady plane shock wave. *Communications on Pure and Applied Mathematics*, **5**(3), pp. 257–300, 1952.

[4] Bird, G.A., *Molecular Gas Dynamics and the Direct Simulation of Gas Flows*. Oxford University Press, 1994.

[5] Reese, J.M., Woods, L.C., Thivet, F.J.P. & Candel, S.M., A second-order description of shock structure. *Journal of Computational Physics*, **117**(2), pp. 240–250, 1995.

[6] Greenshields, C.J. & Reese, J.M., The structure of shock waves as a test of Brenner's modifications to the Navier-Stokes equations. *Journal of Fluid Mechanics*, **580**, pp. 407–429, 2007.

[7] Reddy, M.H.L. & Alam, M., Plane shock waves and Haff's law in a granular gas. *Journal of Fluid Mechanics*, **779**, p. R2, 2015.

[8] Reddy, M.H.L. & Alam, M., Regularized extended-hydrodynamic equations for a rarefied granular gas and the plane shock waves. *Physical Review Fluids*, **5**, p. 044302, 2020.

[9] Von Mises, R., On the thickness of a steady shock wave. *Journal of the Aeronautical Sciences*, **17**(9), pp. 551–554, 1950.

[10] Gilbarg, D. & Paolucci, D., The structure of shock waves in the continuum theory of fluids. *Journal of Rational Mechanics and Analysis*, **2**(5), pp. 617–642, 1953.

[11] Narasimha, R., Das, P. & Lighthill, M.J., A spectral solution of the Boltzmann equation for the infinitely strong shock. *Philosophical Transactions of the Royal Society of London Series A, Mathematical and Physical Sciences*, **330**(1611), pp. 217–252, 1990.

[12] Pham-Van-Diep, G.C., Erwin, D.A. & Muntz, E.P., Testing continuum descriptions of low-Mach-number shock structures. *Journal of Fluid Mechanics*, **232**, pp. 403–413, 1991.

[13] Alsmeyer, H., Density profiles in argon and nitrogen shock waves measured by the absorption of an electron beam. *Journal of Fluid Mechanics*, **74**(3), pp. 497–513, 1976.

[14] Schmidt, B., Electron beam density measurements in shock waves in argon. *Journal of Fluid Mechanics*, **39**(2), pp. 361–373, 1969.

[15] Stokes, G.G., On the theories of the internal friction of fluids in motion, and of the equilibrium and motion of elastic solids. *Transactions of the Cambridge Philosophical Society*, **8**, pp. 287–319, 1845.

[16] Lumpkin, F.E. & Chapman, D.R., Accuracy of the burnett equations for hypersonic real gas flows. *Journal of Thermophysics and Heat Transfer*, **6**(3), pp. 419–425, 1992.

[17] Reddy, M.H.L. & Dadzie, S.K., Reinterpreting shock wave structure predictions using the Navier-Stokes equations. *Shock Waves*, **30**, pp. 513–521, 2020.

[18] Reddy, M.H.L., Dadzie, S.K., Ocone, R., Borg, M.K. & Reese, J.M., Recasting Navier–Stokes equations. *Journal of Physics Communications*, **3**(10), p. 105009, 2019.

[19] Stamatiou, A., Dadzie, S.K. & Reddy, M.H.L., Investigating enhanced mass flow rates in pressure-driven liquid flows in nanotubes. *Journal of Physics Communications*, **3**(12), p. 125012, 2019.

[20] Dadzie, S.K., A thermo-mechanically consistent Burnett regime continuum flow equation without Chapman-Enskog expansion. *Journal of Fluid Mechanics*, **716**, p. R6, 2013.

[21] Brenner, H., Beyond Navier-Stokes. *International Journal of Engineering Science*, **54**, pp. 67–98, 2012.

[22] Dadzie, S.K. & Reddy, M.H.L., Recasting Navier-Stokes equations: Shock wave structure description. *AIP Conference Proceedings*, **2293**(1), p. 050005, 2020.

[23] Paolucci, S. & Paolucci, C., A second-order continuum theory of fluids. *Journal of Fluid Mechanics*, **846**, pp. 686–710, 2018.

Author index

WIT*PRESS* ...for scientists by scientists

Linear and Non-linear Continuum Solid Mechanics

S. HERNÁNDEZ, *University of A Coruña, Spain and* *A. N. FONTAN*, *University of A Coruña, Spain*

Deformable solids, that is to say, those which undergo changes in geometry when subjected to external loads or other types of solicitations, as well as other related topics are the focus of this book.

Within the main field, this text deals with advanced linear elasticity and plasticity approaches and the behavioural study of more complex types of materials. This includes composites of more recent manufacture and others whose material characterisation has only recently been possible. It also describes how linear elastic behaviour extends to anisotropic materials in general and how deformations can result in small or large strain components. The information on plastic behaviour expands to include strain hardening of the materials.

Amongst other new topics incorporated into this volume are studies of hyperelastic materials, which can represent elastomeric and some types of biological materials. A section of the book deals with viscoelastic materials, i.e. those who deform when subjected to long-term loads. The behaviour of viscoplasticity, as well as elasto-viscoplasticity, describes well other types of materials, including those present in many geotechnical sites.

The objective of this volume is to present material that can be used for teaching continuum mechanics to students of mechanical, civil or aeronautical engineering. In order to understand the contents the reader only needs to know linear algebra and differential calculus.

Examples have been included throughout the text and at the end of each chapter, exercises are presented which can be used to check on comprehension of the theoretical information presented.

ISBN: 978-1-78466-271-4   eISBN: 978-1-78466-272-1
Published 2021 / 206pp

www.ingramcontent.com/pod-product-compliance
Lightning Source LLC
Chambersburg PA
CBHW062005190326
41458CB00009B/2978